AF510649

Miscellaneous Antiquities, Number XVIII:

The Strawberry Hill Press & Its Printing House

The Strawberry Hill Press & Its Printing House

AN ACCOUNT AND AN ICONOGRAPHY

Stephen Clarke

adde quod ingenuas didicisse fideliter artes
emollit mores nec sinit esse feros.

Ovid, *Epistulae ex ponto,* II.9.47–48

The Lewis Walpole Library, Yale University · 2011
Distributed by Yale University Press, New Haven and London

Copyright ©2011 by Yale University. All rights reserved.

This book may not be reproduced or transmitted, in whole or in part, including illustrations, in any form (except in reviews for the public press and other copying permitted by Sections 107 and 108 of the U.S. Copyright Law), without permission in writing from the publisher.

Library of Congress Cataloging-in-Publication Data

Clarke, Stephen, 1949–
The Strawberry Hill Press and its printing house : an account and an iconography / Stephen Clarke.
p. cm. — (Miscellaneous antiquities ; XVIII)
Includes bibliographical references and index.
ISBN 978-0-300-17040-5 (alk. paper)
1. Strawberry Hill Press (Twickenham, London, England) 2. Walpole, Horace, 1717–1797. 3. Private presses—England—History—18th century.
4. Twickenham (London, England)—Imprints. I. Title.
Z232.S9C58 2011
070.5'930942—dc22
2010051067

Contents

Foreword

In 1772 Horace Walpole published two issues of *Miscellaneous Antiquities,* both in editions of 525 copies. He intended to satisfy what he gauged to be a general "taste for anecdotes and historic papers, for ancient letters that record affairs of state, illustrate characters of remarkable persons, or preserve the memory of former manners and customs." Walpole's admirable plan was to publish entertaining original manuscripts and printed pieces, "now little known," from his own and his friends' collections. He over-estimated the public's capacity for such amusements, however, as he wrote to his friend William Mason in March 1773: "The *Miscellaneous Antiquities* have not sold above a fifth of them, so there will be no more."

In 1927, as he recalls in *Collector's Progress,* Wilmarth Sheldon Lewis had the temerity to resume Walpole's short-lived series. He was confident that the present taste was "identical with that of 1772" and would embrace antiquities of the sort to be found in his library. As it happened, the first three numbers of Lewis's *Miscellaneous Antiquities* resulted in almost the same disappointing sales as did Walpole's: from editions of 500 copies, only about 130 were sold. Lewis's response was not to abandon the project entirely but to carry on with the series until 1940, privately printing what he calls "odds and ends" from his collection in small editions as Christmas keepsakes.

Given this history, it might have seemed foolhardy to launch a second revival of *Miscellaneous Antiquities* in 2004. But the value to scholarship of primary sources is even clearer now than it was in the 1920s, and the presence in the Lewis Walpole Library of materials that are unknown, or known only to a few scholars, encouraged the present Editor to resume the venerable series. The present study, Stephen Clarke's ambitious reconstruction of the history of Walpole's Printing House through its iconography, is a project that both Walpole and Lewis would surely have endorsed. Meticulously bringing together many disparate textual and visual sources, he illuminates the material history of the building and, in doing so, also tells a compelling story of Walpole's Strawberry Hill Press, its publications, its printers, and its legacy.

Margaret K. Powell
The Lewis Walpole Library

MISCELLANEOUS

ANTIQUITIES;

O R,

A COLLECTION OF

CURIOUS PAPERS:

Either republiſhed from SCARCE TRACTS, or now firſt printed from ORIGINAL MSS.

NUMBER I.

TO BE CONTINUED OCCASIONALLY.

Invenies illic et feſta domeſtica vobis.
Sæpe tibi Pater eſt, ſæpe legendus Avus.

OVID. Faſt. lib. I.

STRAWBERRY-HILL:
PRINTED BY THOMAS KIRGATE, M.DCC.LXXII.

Title page from the first number of Horace Walpole's *Miscellaneous Antiquities*

Miscellaneous Antiquities

The first two numbers were edited by Horace Walpole; numbers 3–16 were edited by W. S. Lewis excepting as noted; number 17 was edited by Matthew J. Kinservik.

1 *Honour Military and Civill.* Strawberry Hill: Printed by Thomas Kirgate, 1772. [525 copies]

2 *The Life of Sir Thomas Wyat, The Elder.* Strawberry Hill: Printed by Thomas Kirgate, 1772. [525 copies]

3 *A Note Book of Horace Walpole.* Mount Vernon, N.Y.: William Edwin Rudge, 1927. [500 copies]

4 *Notes by Lady Louisa Stuart on George Selwyn and His Contemporaries, by John Heneage Jesse.* Edited from the original manuscript by W. S. Lewis. New York: Oxford University Press; London: H. Milford, 1928. [500 copies]

5 *Horace Walpole's Fugitive Verses.* Edited by W. S. Lewis. New York: Oxford University Press; London: H. Milford, 1931. [500 copies]

6 *The Forlorn Printer: Being Notes on Horace Walpole's Alleged Neglect of Thomas Kirgate,* by W. S. Lewis. Walpole Printing Office, 1931. [50 copies]

7 *Anecdotes Told Me by Lady Denbigh,* [by] Horace Walpole. Farmington, Conn.: Privately printed, 1932. [50 copies]

8 *Horace Walpole's Letter from Madame de Sévigné,* by W. S. Lewis. Farmington, Conn.: Privately printed, 1933. [100 copies]

9 *Some Short Observations for the Lady Mary Stanhope Concerning the Writing of Ordinary Letters, by Philip, Second Earl of Chesterfield.* Edited by W. S. Lewis. Farmington, Conn.: Privately printed, 1934. [100 copies]

10 *Le Triomphe de l'Amitié, ou, l'Histoire de Jacqueline et de Jeanneton,* by W. S. Lewis. Farmington, Conn.: Privately printed, 1935. [100 copies]

11 *The Duchess of Portland's Museum,* by Horace Walpole; with an introduction by W. S. Lewis. New York: The Grolier Club, 1936. [450 copies]

12 *Bentley's Designs for Walpole's Fugitive Pieces,* by W. S. Lewis. Farmington, Conn.: Privately printed, 1936. [100 copies]

13 *Memoranda Walpoliana,* with an introduction by W. S. Lewis. Farmington, Conn.: Privately printed, 1937. [100 copies]

14 *Letters to and from Madame du Deffand and Julie de Lespinasse.* Edited by Warren Hunting Smith. New Haven: Yale University Press; London: H. Milford, Oxford University Press, 1938. [500 copies]

15 *The Impenetrable Secret, probably Invented by Horace Walpole.* An Explanation of the Secret. With a Note on the Original by W. S. Lewis. Farmington, Conn.: Privately printed, 1939. [100 copies]

16 *Notes by Horace Walpole on Several Characters of Shakespeare.* Edited by W. S. Lewis. Farmington, Conn.: Privately printed, 1940. [100 copies]

17 *"The Production of a Female Pen": Anna Larpent's Account of the Duchess of Kingston's Bigamy Trial of 1776.* A facsimile edition, transcribed and with an introduction by Matthew J. Kinservik. New Haven, Conn.: The Lewis Walpole Library, Yale University, 2004.

Preface and Acknowledgments

THIS STUDY developed from an initial, almost incidental interest in views of the Printing House of the Strawberry Hill Press. The Press has long been celebrated as possibly the first important private press in England, and the vehicle by which Horace Walpole could circulate his writings and those of his friends, as well as antiquarian material that he found and edited. Its productions were eagerly collected in his lifetime, as they have been ever since, and they have been extensively studied by scholars and bibliographers. Yet investigation of the rich holdings of the Lewis Walpole Library at Yale University produced a surprising number of unfamiliar images.

It also became apparent that despite Paget Toynbee's printing of Walpole's *Journal of the Printing-Office at Strawberry Hill* (1923), with its rich store of notes, and Allen Hazen's magisterial *Bibliography of the Strawberry Hill Press* (1942), there was no sustained narrative history of the Press. The essence of the Press was the variety of its publications—from Gray's *Odes* and that foundation stone of English art history, the *Anecdotes of Painting in England,* to the slightest of ephemera—its longevity, and the survival of records of its activities in the *Journal of the Printing-Office* and in Walpole's correspondence. These reveal the varied uses of the Press, and the extent to which it became an extension of Walpole's cultural endeavor as antiquarian, art historian, man of letters, and representative of a privileged literary and social circle. The Press was central to this, and it is no coincidence that Walpole's best known and longest-serving printer, Thomas Kirgate, evolved into his secretary and amanuensis.

I have therefore attempted to provide an account of the Press and its achievement, and to put this in the context of the unexpected number of surviving pictures of the Printing House, and of Kirgate as its most celebrated printer. This has involved a consideration of the activities of those collectors who were also extra-illustrators of books (a fashion in whose creation Walpole was closely involved), and also of the artists who, in the context of Strawberry Hill, worked to meet the demand for extra-illustration. I have used maps to show how the Printing House fitted into Walpole's estate at Strawberry Hill and, in conjunction with later prints and photographs, to record the unfamiliar story of the building's decay and adaptation after Walpole's death and Kirgate's departure.

It will come as no surprise to anyone familiar with Walpole studies that this book firmly grounded in the resources of the Lewis Walpole Library. Readers who make their way to the concluding iconographical catalogue will see that the vast majority of prints, drawings, and maps listed now reside there, assembled with awe-inspiring comprehensiveness and dedication by Wilmarth Sheldon Lewis, to whose scholarly and collecting achievement generations of researchers are indebted. My thanks go in particular to Margaret Powell as Librarian for her support and enthusiasm for this project; to her colleagues Cynthia Roman and Susan Odell Walker, both of whom gave freely of their time and knowledge, and Brian Parker, who has been most helpful in providing technical advice and the illustrations of items at the library; and to all the Lewis Walpole Library staff, for providing the most supportive and congenial research library facilities that any scholar could reasonably ask for.

I have been very privileged to benefit from the skill and sureness of eye of John Gambell and Lesley Baier at the Office of the Yale University Printer and am grateful to them for their commitment ever since I first showed them some photocopy images and tried to explain the story that I thought those pictures could tell.

Special thanks go to Peter Sabor for reading and making suggestions on a draft of the text, and for numerous discussions on Walpolian and related matters. On issues of printing history, David Chambers has been particularly helpful, and Mirjam Foot, Henry Woudhuysen, and John Gretton also provided much valued additional advice and information. Michael Snodin made available his formidable understanding of the history of the house and estate at Strawberry Hill, while Kevin Rogers of Inskip and Jenkins, architects, who have recently carried out the triumphant restoration of the house for the Strawberry Hill Trust, drew the plans that show the various transformations in the layout of the Printing House. I must also thank Alexander O. Trotter, who undertook the essential labor of the Index. The Earl Waldegrave, Lord Waldegrave of North Hill, the Hon. Nino Strachey, Sheila O'Connell at the British Museum, Alan Jutzi and Stephen Tabor at the Huntington Library, Rachel Bond at Eton College Library, Jenni Rodda of the Institute of Fine Arts at New York University, and Christopher Hunwick as Archivist to the Northumberland Estates all made material in their collections available and answered queries about it. Others who have kindly allowed items held by them to be reproduced are the Richmond upon Thames Local Studies Library and the Friends of Strawberry Hill. I am also grateful to Kelvin Freeman, who photographed four of the images, and to Christine Jarvis and Father Gerard Devlin of St Mary's University College at Strawberry Hill, who respectively provided photographs and access.

But my greatest thanks must, as always, go to my wife, Sally, for unvarying and constant support; to her I would wish to dedicate this book.

The Strawberry Hill Press & Its Printing House

1

William Watts, *"View of Strawberry Hill" from the East,* ca. 1780s. Watercolor, 13 x 18.3 cm (image), laid on wash mount

Unless otherwise indicated, all works illustrated in this publication are from the collection of the Lewis Walpole Library, Yale University.

"My abbey is a perfect college or academy"

June 25th [1757] *I erected a printing-press at my house at Strawberry Hill. Aug. 8th published two odes by Mr Gray, the first production of my press.*[1]

WITH THESE TWO ENTRIES from his autobiographical account known as "Short Notes of the Life of Horatio Walpole," Horace Walpole described the genesis of perhaps the most famous private press in England. Ten years after he had first rented the former coachman's cottage at Strawberry Hill in Twickenham, and having already purchased the property and carried out highly original work to extend and convert it into a Gothic castle in miniature (Fig. 1), Walpole decided to set up his own press. His enthusiasm is best captured in two letters written in the summer of 1757. On 12 July he wrote to John Chute:

> On Monday next the Officina Arbuteana opens in form. The Stationers' Company, that is, Mr Dodsley, Mr Tonson, etc. are summoned to meet here on Sunday night. And with what do you think we open? *Cedite,* Romani *Impressores*—with nothing under Graii *Carmina.* I found him in town last week: he had brought his two odes to be printed. I snatched them out of Dodsley's hands, and they are to be the first-fruits of my press. An edition of Hentznerus, with a version by Mr Bentley and a little preface of mine, were prepared, but are to wait... I tell you no news, for I know none, think of none. Elzevir, Aldus and Stephens [i.e., Estienne], are the freshest personages in my memory.[2]

Walpole's excitement was not entirely shared by the cautious and retiring Thomas Gray at being obliged to consent to his *Odes* becoming Walpole's first exercise in printing (see Fig. 5).[3] One might have expected to find Robert Dodsley (1703–1764), as publisher of the *Odes,* attending Strawberry Hill to inspect the fledgling printing operation, but it is interesting that Walpole was also able to summon another prominent bookseller, Jacob Tonson (1714–1767). It is also noteworthy that the *Odes* displaced Paul Hentzner's *Journey into England in the Year 1598,* which was dedicated by Walpole to the Society of Antiquaries and was the first of many antiquarian offerings from the Press.

The second letter, to Horace Mann, followed some three weeks later:

> Shall I tell you what, more than distance, has thrown me out of attention to news? A little packet which I shall give your brother for you, will explain it. In short, I am turned printer, and have converted a little cottage here into a printing-office—My abbey is a perfect college or academy—I keep a painter in the house and a printer—not to mention Mr Bentley who is an academy himself. I send you two copies (one for Dr Cocchi) of a very honourable opening of my press—two amazing odes of Mr Gray—they are Greek, they are Pindaric, they are sublime—consequently I fear a little obscure.[4]

Here Walpole expounds his perception of Strawberry Hill as a center from which to dispense patronage of the arts: a rural retreat, away from the bustle of society and the court, where he could surround himself with Richard Bentley (1708–1782), architectural designer and graphic artist; the Swiss painter Johann Heinrich Müntz (1727–1798); and now his own printer, William Robinson. Both Müntz's and Bentley's primary function was to adorn Strawberry Hill: Müntz with his paintings of the house and grounds, Bentley with Gothic designs, extravagant and improbable fireplaces, and the hall and staircase, "the most particular and chief beauty of the castle."[5] Just how striking and unexpected that hall and staircase were, with a balustrade echoing an original from Rouen Cathedral and walls painted to imitate low relief, the theatrical ensemble lit by a single candle in the Gothic lantern designed by Bentley, is captured in Lady Townshend's exclamation when mounting the stairs—"Lord God! Jesus! What a house!"[6]

The role of the printer, meanwhile, was to provide additional luster and diffuse the productions of the "perfect college or academy" to the wider world. Two months after the Press was erected, Bentley wrote and Müntz set a sonnet, "To the Printing-Press at Strawberry-Hill," which they printed and presented to Walpole, expressing these aspirations:

> Yet here shalt thou let Dullness flow
> On the blank page no more;
> But like Preferment, Sense bestow
> Where there was none before.[7]

The verses proceed to question the uneven reception of Gray's *Odes* in comparison to the popular success of Samuel Richardson's novels, underscoring the potential of the Press to bring original material to the public. In reality, much of that material was to consist of the antiquarian and other works that Walpole wrote or edited, as well as pieces written by or for the benefit of his friends. Antony Griffiths has noted that Walpole was above all an antiquarian, collecting manuscripts, historical fragments, and engraved historical portraits.[8] He was also at the center of an elite circle of male sociability, extending from antiquaries such as the Reverend William Cole

(1714–1782), through poets such as Thomas Gray (1716–1771) and the Reverend William Mason (1725–1797), virtuosi such as John Chute (1701–1776), diplomats such as Sir Horace Mann (1706–1786), and soldiers and statesmen such as his cousin General Conway (1719–1795), to socialites such as Gilly Williams (ca. 1719–1805) and George Selwyn (1719–1791), and gentlemen of leisure such as George Montagu (ca. 1713–1780). If this at first appears a disparate group, it is bound together by a number of common threads: assured social status, historical awareness, and the centrality of taste. George Haggerty has written of Walpole's building and collecting enterprise at Strawberry Hill as a collaborative project in which he implicated his most intimate friends.[9] While the initial core of this collaboration was the Strawberry Committee of Walpole, Bentley, and Chute, who provided the designs for the building works, the circle rippled outward, recorded in and bound together by the network of Walpole's extensive correspondence, a number of his friends and correspondents identifying in various ways with his building projects, and some of them providing him with historical relics to add to his collection.[10]

One example of a creative exercise undertaken by Walpole in conjunction with some of his circle is the extra-illustration of books, and his account of this activity, in a letter to William Mason of 7 May 1775, provides a revealing perspective on his own ambitions and aspirations and on the role of the Strawberry Hill Press in realizing them:

> I who, simpleton as I was, loved to be an author, am so ashamed of my own stuff, and so convinced that nobody but you and Gray could write, have taken shame to myself, and forsworn the press; yet as I cannot be idle, it is impossible. I have invented a new and very harmless way of *making books,* which diverts me as well, and brings me to no disgrace. I have just made a *new book,* which costs me only money, which I don't value, and time which I love to employ. It is a volume of etchings by *noble authors.* They are bound in robes of crimson and gold; the titles are printed at my own press, and the pasting is *by my own hand.*[11]

To help untangle the variety of messages contained in this passage, three points need to be made. First, there is throughout Walpole's life a continuing tension between his desire for celebrity and his fear of ridicule. Gilly Williams's often quoted summary—"*Monstrari digito praetereuntium* has been his whole aim. For this he has wrote, printed, and built"—aptly captures the way in which Walpole as builder, collector, author, and printer sought the public gaze.[12] But public exposure inevitably brought with it the risk of public criticism, and the cultured and ingenious Mr. Walpole did not relish being criticized by the antiquaries for his misuse of evidence in his defense of King Richard III, still less being traduced for his alleged ungenerous behavior to the young Chatterton, and still less again being abused as a hermaphrodite infatuated with his cousin Conway in William Guthrie's scurrilous pamphlet *A Reply to the Counter Address.*[13]

Second, Walpole's persistent use of self-deprecation, a recurrent trope in his correspondence, is a literary device to deflect the criticism that his desire "to be pointed out by the fingers of those passing by" attracted. The shame he affects in this passage in respect to his own writings should in no sense be taken at face value; Walpole started (and then aborted) an edition of his own works in 1770, but he made sure that his young friend Mary Berry (1763–1852) was able to produce a complete edition of his works in five substantial quarto volumes within fifteen months of his death. Walpole may have affected the negligent air of the gentleman, what Morris Brownell has characterized as *sprezzatura,* but this does not in any way undermine the seriousness of his cultural and antiquarian endeavors.[14]

Third, there is Walpole's exaggerated view of the achievements of his friends, the naturally gifted amateurs of good birth who were able almost without effort to write or draw or sculpt unhampered by the mundane drudgery of any formal training. This affected both his assessment of their merits and the productions of his press. Not only did he overpraise the poetry of the now largely unread Mason (who was at least an acknowledged and published poet), but he was also happy to describe Conway's daughter, the sculptress Anne Seymour Damer (1748–1828), as modeling like Bernini, and to mention his protégée Lady Diana Beauclerk (1734–1808) in the same sentence as Michelangelo, as well as praising the astonishing genius of Lady Lucan (d. 1814) in copying pictures, a talent brought to marvelous perfection in a modest five months.[15] This is partly an issue of snobbery; as his most sensitive biographer has commented, Walpole "could never quite forget, however familiarly he met them in social life, that Reynolds was a painter and Garrick a player."[16] But as Cynthia Roman has shown in relation to the amateur artists he promoted as examples of female genius, it was also a matter of using conscious hyperbole as a strategy for engaging in a critical dialogue with the male and professional world of the Royal Academy, just as the gentlemen writers who had their trifles printed at the Strawberry Hill Press were praised in part because they represented the antithesis of commercial authorship.[17] Walpole's role as patron of the arts is compromised by the dichotomy between his apparently excessive praise for the amateur ladies and gentlemen of talent, and his rather more difficult relationships with professional artists; the three original members of the "perfect college or academy," Bentley, Müntz, and the printer Robinson, were all in due course turned away, and other artists from whom Walpole was subsequently to distance himself included the draftsman and designer Edward Edwards (1738–1806) and, though he was never in any sense a dependent, the architect Robert Adam (1728–1792). As for the engravers he employed, Walpole complained that they were "so extravagant, though so indifferent, that I have almost given over the pleasure of having engravings."[18] For a jaundiced account of Walpole's limitations as patron and host,

2
D.P. Pariset, after Pierre-Etienne Falconet, *"The Revd. Mr. James Granger,"* 1768. Stipple engraving, 17.5 x 12.4 cm (image), 19.6 x 14.5 cm (plate)

3
J. Harding, after John Nixon, *"Anthony Storer Esqr at the Sale of Mr Aytons (the Banker's) Prints,"* 1795. Etching, 13 x 9 cm (sheet). The Provost and Fellows of Eton College

there is none more embittered than that of William Beloe (1758–1817), the former under-librarian of the British Museum, who complained that

> As far as verbal communication went, or communication of what he retained in his memory, availed, all his stores were at the service of literary men, and many of our modern popular books owe much of their zest and interest to this sort of assistance received from Lord [Orford].... Further than such communication, with perhaps the exception of a scanty dinner at Strawberry-hill, there is no instance on record of his liberality having proceeded. He certainly was proud of being considered as a sort of patron of literature, and a friend to literary men, but he did not choose to purchase the pre-eminence at a higher price than a little flattery and praise, and a pudding neither over large nor over solid.[19]

With these three general points in mind, Walpole's letter to Mason disavows (predictably inaccurately) his future use of his press to publish his own modest works and celebrates his binding of a collection of etchings by noble amateurs, with a title page from his own press, while underscoring his leisured status by stressing his freedom to use his time as he wishes and his lack of concern at the cost. Lucy Peltz has pointed out in an illuminating study that the rage for extra-illustration and the minor industry of print collecting and publishing required to satisfy it began in the elite group of Walpole's circle, producing or creating a book as a luxury and unique item. The Reverend James Granger (ca. 1723–1776; Fig. 2), after whom the practice of grangerizing is named, was a correspondent, as were Anthony Storer (1746–1799; Fig. 3) and Richard Bull (1721–1805; Fig. 4), both of whom compiled extensive collections of extra-illustrated copies of Strawberry Hill books; Peltz records that Bull owned at least twenty-five Strawberry Hill Press books, of which he extra-illustrated sixteen.[20]

4
Arthur Devis, *Portrait of Mr. and Mrs. Richard Bull,* 1747. Oil on canvas, 106.7 x 86.4 cm. New York University Collection, Conservation Center, cc 91.70

Walpole was suitably flattered by such attention. In 1781 he wrote to Cole that "Mr Bull is honouring me, at least my *Anecdotes of Painting,* exceedingly. He has let every page into a pompous sheet, and is adding every print of portrait, building, etc., that I mention and that he can get, and specimens of all our engravers. It will make eight magnificent folios [in fact fourteen], and be a most valuable body of our arts."[21] Storer, whose conversion from macaroni to connoisseur Walpole noted with surprise, was with Bull to become "among the half dozen principal collectors in England of engraved portraits."[22] He is shown in caricature in the role of discriminating, possibly fastidious, collector, engrossed at a print sale. In contrast, the marriage portrait of Bull and his wife by Arthur Devis is an exposition of polite society and its values; the relaxed poses of the couple may owe something to French engravings, while behind them the ceramics on the mantelpiece and the classical landscapes and busts on the wall bespeak their taste—the two images showing the two faces of connoisseurship.[23] Walpole's correspondence with Bull consists of an exchange of prints, drawings, and compliments. It was Bull who compiled one of the richest of all the extra-illustrated copies of the *Description* of Strawberry Hill, using principally the artist and architectural draftsman John Carter (1748–1817), and Bull lent it to the elderly Walpole, who in February 1792 returned it with the following letter:

> I am exceedingly obliged to you for trusting me so long with the book of Strawberry, to which you have done so *much honour,* & I fear at *too great expence.*
>
> I have answered two or three of your queries with a pencil slightly. The genealogic table I have put into the book.
>
> I am sorry Carter cannot draw heads as well as buildings; I cannot commend his Hen. 7. & Caligula. the bust on the last leaf with the latter is unknown, & not a good one, & is actually one of the plates of the Antiquit. Middleton.
>
> The print from my Rembrandt is abominably coloured, & has a red mitre instead of a white one.
>
> I am sorry you cannot let me see you oftener; I shd very willingly exert my voice a little to tell you how much I am
>
> dear Sr
> yr exceedingly gratefull humble sert
> Orford
>
> P.S. Carter has called the *Simpulum* in the Library a *ladle.*[24]

The combination of courtesy and antiquarian concern for accuracy is typical of their correspondence. And it is these concerns of informed and committed antiquarianism combined with the beautiful negligence of a gentleman, of an engaged social group untroubled by the harsh realities of trade, of the politeness of social exchange but also of the celebration of genius, all within the welcoming shadow of Walpole's extraordinary house and collection, that form the setting for the establishment of the Strawberry Hill Press.

Walpole recorded the activities of the Press not merely in his correspondence, but also in a separate journal, the *Journal of the Printing-Office at Strawberry Hill,* which was edited by Paget Toynbee in 1923.[25] On the inside of the front cover of a quarto notebook, Walpole wrote beneath his name and the date 1757 the words "Archbishop Parker kept in his house a Painter, Engraver, and Printer." This reflects the fact that after William Robinson's arrival in June 1757, Walpole was keeping in his own house a painter, a designer, and a printer, but it also underscores Walpole's awareness that there had been an example of a private printing press some two hundred years earlier. At Lambeth Palace in 1573, that patron of scholars Archbishop Parker could write to Lord Burghley that "I have within my home in wages, drawers and cutters, painters, limners, writers, and bookbinders."[26] Closer to Walpole's time, in 1739, the Reverend Francis Blomefield (1705–1752) had printed the early parts of his *Essay towards a Topographical History of the County of Norfolk* at his own press at his rectory at Fersfield. Then there was the notorious but short-lived example of the press set up by John Wilkes (1725–1797) at his house in Westminster in 1763, in order that he could reprint *The North Briton* in octavo format. Walpole had no respect for Wilkes, whom he dismissed as "unrestrained either in his conduct or conversation," and had always insisted that his own press should not be polluted with politics ("I vowed and declared it never should be employed either in politics or satire," he told William Mason), but there are two unlikely bonds that link the press at Strawberry Hill and Wilkes's press at Great George Street.[27] The discovery of Wilkes as joint author of the infamous *Essay on Woman* was initiated by Walpole's former printer Thomas Farmer, when over lunch he picked up a proof sheet in which one of Wilkes's printers had wrapped his food; and when in April 1763, in the middle of the furor over *The North Briton* no. 45, Wilkes published no. 46, he had it inscribed with a fine sense of irony as "Printed at Strawberry Hill. 1763."[28]

Far more long-lasting was the Grange Press at Darlington, set up by the lawyer and antiquary George Allan (1736–1800) and active from about 1768 for some thirty years, printing primarily antiquarian and historical material, with some ephemera. Unlike Walpole, Allan set his own type and undertook some of his own printing on a common press installed in his house, generally producing no more than fifty copies of each item.[29] There were also a number of minor presses set up in the last quarter of the century, at places such as Glynde and Bishopstone in Sussex and Madeley in Shropshire, mostly established by clergymen to print religious works and poetry: Bishopstone was the press of the poet the Reverend James Hurdis (1763–1801). But what immediately distinguishes the press at Strawberry Hill from all these is its variety, the importance of some of its imprints, its longevity, the attention it has always received from collectors, and the detailed records of it that survive in word and image.

A.T. Hazen's *Bibliography of the Strawberry Hill Press* (1942) lists some thirty-four books and pamphlets printed at the Press, and forty-eight "detached pieces"—mostly poetry, but including the rules by which visitors could obtain a ticket to be shown the house, and the label for the pedestal of the china tub immortalized in Gray's "Ode on the Death of a Favourite Cat."[30] There were also title pages for books that Walpole had extra-illustrated or assembled, such as the title page for the volume of etchings by noble amateurs noted above, and the title pages for his collections of poems, plays, and tracts from the reign of George III; there was even a title page for the collection of loose or detached pieces printed at the Press. Hazen's thirty-four books and pamphlets can be broken down by content as follows: nine antiquarian and historical titles, of which Walpole wrote four and edited the others; eight productions of friends, mostly poetry; six items on Strawberry Hill, cataloguing its contents; six titles of Walpole's other writings; three books printed for charitable purposes (one poetry, one historical memoirs, one biographical account); one edition of classical literature; and one catalogue of a collection of paintings. If one turns to the detached pieces, the vast majority are *vers de société* and poetic trifles. In all, the overwhelming themes are antiquarianism, the production of Walpole's own writings, the recording of Strawberry Hill, and elite social exchange. Walpole's use of the Press to print his own works (of Hazen's total of thirty-four books, half were written by Walpole) underscores a primary function of the Press, in that it gave Walpole control of circulation; he could decide whether to advertise titles and have them distributed as widely as possible through the London booksellers (such as the *Anecdotes of Painting,* for which Paget Toynbee reproduces an advertisement for the first two volumes, which could be bought from Walpole's bookseller William Bathoe), or print tiny numbers (he recorded that he printed fifty copies of *The Mysterious Mother,* and only six of his *Hieroglyphic Tales*) and then determine with how exclusive and limited a circle they should be shared.[31] This flexible system could sidestep those inconveniences of authorship that he found so uncongenial—being attacked for one's views in pamphlets and abused by "those dirt-carts and their drivers," the magazines.[32]

In terms of substance, the most important item as literature is unquestionably the first fruit of the Press, Gray's *Odes,* seized by Walpole in order that he could christen his Press with a friend's poetry of obvious merit that had the added attraction of publicizing his new venture (Fig. 5).[33] The other poems and plays printed at the Press (excluding for the moment those written by Walpole) were of a very different order; Countess Temple's *Poems* (1764), the Honorable Richard Fitzpatrick's *Dorinda, A Town Eclogue* (1775), and the last pamphlet issued from the Press, Hannah More's *Bishop Bonner's Ghost* (1789), shade imperceptibly into the verse trifles that constitute so many of the detached pieces. The gallantry that Walpole could bring to those perishable lines might be thought to supplement their negligent, insubstantial elegance; as Walpole wrote to Hannah More (1745–1833) in July 1789 of printing *Bishop Bonner's Ghost,* "My press can confer no honour; but when I offer it, it is a certain mark of my sincerity and esteem. It has been dedicated to friendship, to charity—too often to worthless self-love; sometimes to the rarity of the pieces, and sometimes to the merit of them—now it will unite the first motive and the last."[34] But Walpole, that "witty, sarcastic, ingenious, deeply-thinking, highly-cultivated, quaint, though evermore gallant and romantic, though very mundane, old bachelor of other days" (in the words of

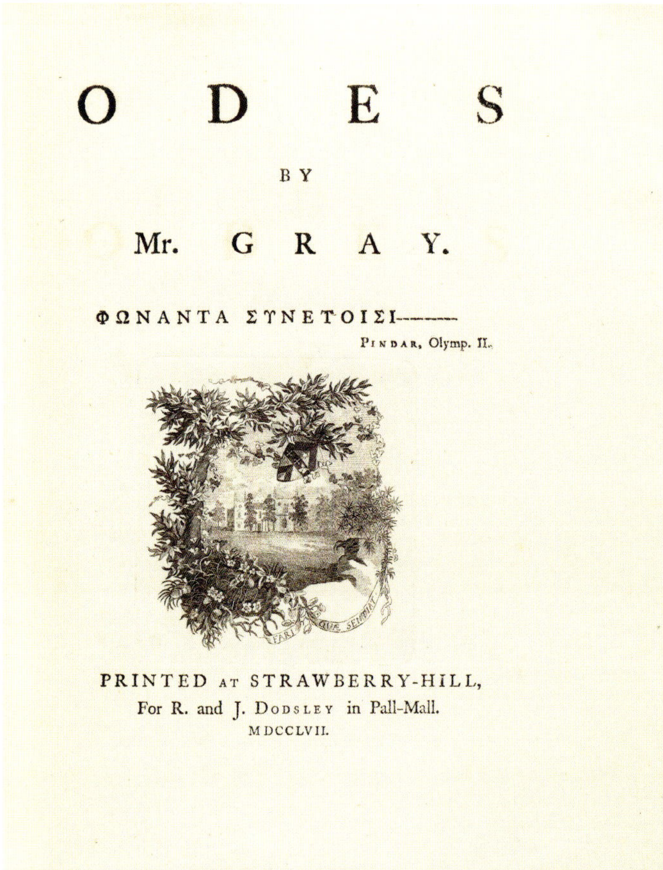

ODES

BY

Mr. GRAY.

ΦΩΝΑΝΤΑ ΣΥΝΕΤΟΙΣΙ——
PINDAR, Olymp. II.

PRINTED AT STRAWBERRY-HILL,
For R. and J. DODSLEY in Pall-Mall.
MDCCLVII.

5
Title page, *Odes by Mr. Gray.* Strawberry Hill Press, 1757

Frances Burney), was no Sternean sentimentalist.[35] In the same month in which he had written to the pious More, he wrote to Edward Jerningham (1737–1812) that his long-standing printer, Thomas Kirgate (1734/5–1810), "has just delivered the virgin Hannah of a divine babe, of which she fell in labour on fancying she saw a ghost."[36] But gallantry could descend into fulsome flattery, as in Walpole's request to Lord Harcourt (1736–1809), an amateur etcher (with whom he subsequently fell out), to be allowed to print some poetry of his wife:

> Oh! that ever I might be honoured so far, as to join certain lines [poems by the Countess] to those of Lady Temple. The editions of Strawberry would be immortal, and Cipriani should design a frontispiece in which Friendship should present the sister-poetesses to Apollo—and the best engraver in England should etch it—no, my Lord, not Bartolozzi, but an idle creature, as humble as his wife... [that is, Harcourt himself].[37]

Turning from this obeisance to the antiquarian items, Walpole's *Anecdotes of Painting* (1762–1771, but the last volume not distributed until 1780)—the ordering and reduction of the chaotic, voluminous manuscripts of George Vertue (1684–1756) to five carefully arranged volumes recording the painters, architects, sculptors, and engravers who had worked in England—was perhaps the most influential production of the Press, not only for its dissemination of Vertue's materials, but also for Walpole's commentary. There were numerous editions and reprintings as late as 1888, with additional material issued in the form of a supplement as recently as 1937.[38] Before the Walpole Society (founded in 1911 to publish original art historical scholarship) published Vertue's manuscripts, Lionel Cust wrote that "an examination of the actual manuscripts themselves will justify in full the extraordinary importance of the work achieved by Horace Walpole"; and Lawrence Lipking has added that, despite the absence of any real conceptual or theoretical rigor among the antiquarian display of fact and anecdote, Walpole's work "shaped the history of English art... The Walpole Society is not idly named."[39] It was if anything these volumes that earned Walpole the sobriquet of the father of English *virtù*.

The form that Walpole used to order Vertue's material, a series of short biographical accounts of the artists in chronological order, was repeated in his *Catalogue of the Royal and Noble Authors of England* (1758; Fig. 6), an inquiry consistent with Walpole's interests as the "lifelong collector of portraits, which he valued for the sake of anecdote rather than art"—or, as he explained in relation to his political memoirs (but making allowance for his assumed modesty), "I am no historian: I write casual memoirs, I draw characters; I preserve anecdotes, which my superiors the historians of Britain, may enchase into their weighty annals, or pass over at their pleasure."[40] The great Walpole collector and scholar Wilmarth Sheldon Lewis noted that Walpole devoted one press, or bay, of his library at Strawberry Hill to English royal and noble

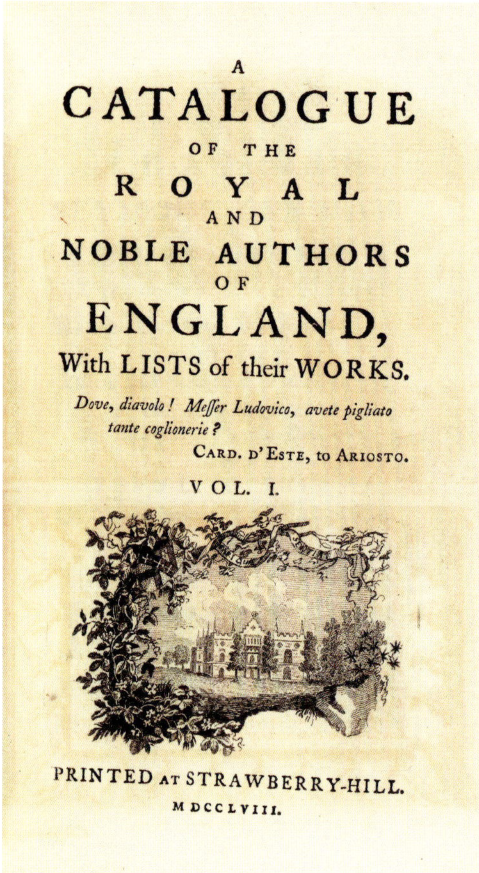
A
CATALOGUE
OF THE
ROYAL
AND
NOBLE AUTHORS
OF
ENGLAND,
With LISTS of their WORKS.
Dove, diavolo! Messer Ludovico, avete pigliato tante coglionerie?
CARD. D'ESTE, to ARIOSTO.
VOL. I.
PRINTED AT STRAWBERRY-HILL.
MDCCLVIII.

6
Horace Walpole, Title page, *A Catalogue of the Royal and Noble Authors of England, With Lists of their Works.* Strawberry Hill Press, 1758

authors, whose contribution to the history of polite literature the *Catalogue* celebrated.[41] The catalogue format also underpins one of the most important productions of the Press, Walpole's *A Description of the Villa of Mr. Horace Walpole, Youngest Son of Sir Robert Walpole Earl of Orford, at Strawberry-Hill near Twickenham, Middlesex. With an Inventory of the Furniture, Pictures, Curiosities, &c.* This was originally printed in 1774 (the title omitting the "Mr.") in an edition of one hundred copies, with an expanded edition of two hundred copies printed in 1784 and supplied with engravings. It is a typically Walpolian touch that the title should celebrate the father whose memory he revered, and to whose great collection of Old Masters the very different collection at Strawberry Hill stood in such stark contrast. The book has a complex printing history. Both editions were supplemented with a series of appendices and lists of additional acquisitions, Walpole continuing to add supplements to the 1774 edition even after the 1784 edition was printed.[42] It is very much a catalogue, listing his collection in great detail, rather than a guide book; it was not available to be used by the many visitors to the house, nor was it formally published or even circulated in

Walpole's lifetime, though a very few copies were given to close friends. His intention was to bequeath copies of the 1784 edition after his death, and he drew up a list of eighty people to whom he wanted copies given, a list he revised shortly before he died in 1797.[43]

The function of the *Description* was accordingly as a record of Walpole's creation at Strawberry Hill, with a view—as explained in the Preface—to the future dispersion of the pictures and rarities. It is the central example of Walpole's use of the Press to commemorate the house and collection (Walpole also printed in 1760 a *Catalogue of Pictures and Drawings in the Holbein-Chamber* and in 1773–74 three variants or offprints of parts of the main text of the *Description*), and it has proved spectacularly successful in that aim. Long after the collection was scattered by the sale of 1842, the *Description* has remained, with George Robins's Sale Catalogue, the key source for recording, tracing, and reassessing the importance of Walpole's collecting achievement at Strawberry Hill.[44] That recording function reached its apogee in the extra-illustrated and expanded copies of the *Description* created by Walpole; by the collectors Bull, Storer, and Charles Bedford (1742–1814); by the artists John Carter and William Bawtree (1744–1824); and by Kirgate himself. Their activities have been discussed by Nicolas Barker in terms of their own extra-illustrating and print collecting pursuits, and in relation to Granger's *Biographical History* and the print collecting of James West (1703–1772). In considering Walpole's own extensively extra-illustrated copy, Ruth Mack has noted how it not only evokes the extraordinary richness and variety of Walpole's collection, but also at the same time recreates the "paper Fabric" of Strawberry Hill on paper—a collection of a collection.[45]

The house spoke of Walpole's antiquarianism, a central element of which (as illustrated in the *Anecdotes of Painting* and the *Catalogue of Royal and Noble Authors,* and equally in the descriptive labels Walpole supplied to some of the items in his collection) was the need to preserve anecdotes and incidental information. It has long been known that the Press was used to provide a label for the bowl on which the pensive Selima reclined, as celebrated in Gray's "Ode on the Death of a Favourite Cat," but it has only recently been discovered that the sculpture of a shouting male head from the Star Chamber, believed by Walpole to be by Torregiano and to represent Henry VII in his death agony, still bears the remains of a printed label, "H.../Torreg"; this label, and some recently identified printed letters and numbers contained in a folded sheet inscribed in Kirgate's hand "Initials for the Bookcases," were quite possibly provided by the Press.[46] The centrality of anecdote and the need to supplement and preserve incidental information were demonstrated in the marginalia Walpole added to many of his books—filling in a missing name here, adding a telling vignette there—and also fired the antiquarian and historical productions of his printing house.[47] Hentzner's *Journey into England* (1757, the second production of the Press), Lord Whitworth's

Account of Russia as it was in the Year 1710 (1758), the picaresque *Life of Edward Lord Herbert of Cherbury* (1764, described by Walpole as "a most delectable work... It is the life of the famous Lord Herbert of Cherbury, written by himself—one of the most curious pieces my eyes ever beheld"), and the Strawberry edition of Walpole's favorite, the *Mémoires du Comte de Grammont* (1772), all fulfilled these criteria.[48] In addition to these publications, Walpole reprinted old tracts and manuscripts, such as the *Copies of Seven Original Letters from King Edward VI. to Barnaby Fitz-patrick* (1772) and the two issues of *Miscellaneous Antiquities* (both also 1772), but he found that there was a limited market for strictly antiquarian pamphlets.

Walpole expressed confidence that he understood the workings of the trade and of potential demand, when setting out to the husband of his correspondent Lady Ossory the possible alternatives for printing the manuscript *Letters* of King Edward VI in Lord Ossory's possession:

> I must ask your Lordship in what manner you would have me print them; I mean, whether for publication, or a smaller number only to give away. ...a smaller number will keep them a curiosity, and yet be sufficient to preserve them. If you like this method, I will print you what number you please, and will send you two or three hundred, and will ask your leave to keep an hundred for myself, as I did for Lord Powis [the owner of the manuscript of the *Life* of Lord Herbert of Cherbury]. He had one hundred copies and I the same—and in two years one copy was sold at an auction for four guineas—you see I have learnt the mysteries of my trade.[49]

He printed two hundred copies of King Edward's *Letters,* but was then disappointed when he expanded that to five hundred copies for sale of the *Miscellaneous Antiquities.* As he wrote to William Cole the following year, "the continuation of the *Miscellaneous Antiquities* is uncertain. I thought the affectation of loving veteran anecdotes was so vigorous, that I ventured to print 500 copies. 130 only are sold—I cannot afford to make the town perpetual presents... All I desire is to pay the expenses, which I can afford much less than my idle moments."[50] This was not, however, Walpole's usual experience; it is far more common to find him concerned at the machinations of the trade, and the inflated prices paid by collectors for Strawberry Hill imprints. In 1764 he complained to the Scottish judge and antiquary Sir David Dalrymple (1726–1792):

> The London booksellers play me all manner of tricks. If I do not allow them ridiculous profit, they will do nothing to promote the sale; and when I do, they buy up the impression, and sell it at an advanced price before my face. This is the case of my first two volumes of *Anecdotes,* for which people have been made to pay half a guinea and a guinea more than the advertised price. In truth the plague I have had in every shape, with my own printers, engravers, the booksellers, etc., besides my own trouble, have almost discouraged me from what I took up at first as an amusement, but which has produced very little of it.[51]

The cost of a set of the *Anecdotes of Painting* continued to soar, exacerbated by the scarcity of complete sets, given the eighteen-year gap between the publication of the first and last volumes. In 1786 Walpole protested to his cousin Thomas Walpole the younger (1755–1840), "My Strawberry editions, merely from their scarcity, sell ridiculously dear. I was forced to pay thirteen guineas yesterday for the *Anecdotes of Painting* in quarto for the King of Poland, who had sent to me for them, and I had not a spare copy left. It is hard to pay almost treble the value for my own impressions!"[52] But then the imprints of the Press always attracted collectors. In addition to their scarcity, they were handsomely presented and cleanly laid out; as his correspondent George Montagu wrote to Walpole in August 1764 on receipt of a copy of *The Life of Edward Lord Herbert of Cherbury,* "Your press every year is brought to bed of some lovely offspring, but this is the prettiest child of all." Walpole may have found less helpful the suggestion of Sir David Dalrymple that the books of the Press should be issued in one uniform size.[53] There is, however, no evidence of any design input by Walpole himself, and the typographical standards to which the Press aspired were not dramatically higher than those of the trade of the time. Compared to Baskerville or Foulis, Strawberry Hill titles, for all their charm and handsomely engraved vignettes and good margins, are not inherently greatly distinguished, and David Chambers has identified bad inking and poor imposition as recurrent problems (see Figs. 16 and 17).[54] But during Kirgate's long tenure as printer, the Press did maintain a consistency of quality of production, and Kirgate was not averse to experimentation; there is at the Lewis Walpole Library at Yale University a version of the large rectangular vignette view of Strawberry Hill (used by the Press from 1764) printed by Kirgate in two colors (Fig. 7), inscribed by him in pencil, "An Attempt to Print in two Colours, first Red, then Black, which I could not perform to my satisfaction."[55]

A related issue was Walpole's limitations as a proofreader, "the most tiresome occupation either as editor or printer."[56] Gray was prompt to point out "some little inaccuracies" in the printing of his *Odes,* a phrase repeated exactly by Conway in relation to Whitworth's *Account of Russia.*[57] Walpole the dilettante presumed to make a virtue of this: as he wrote to Mason (from whom as Gray's literary executor he had hoped to secure some Gray material to print),

> I have not the patience necessary for correcting the press. Gray was forever reproaching me with it, and in one of the letters I have just turned over, he says, "Pray send me the proof sheets to correct, for you know you are not capable of it." It is very true, and I hope future edition-mongers will say of those of Strawberry Hill, they have all the beautiful negligence of a gentleman.[58]

The esteem in which Strawberry Hill editions have always been held suggests that his ambition has been fulfilled.

Walpole sometimes addressed his correspondents in the guise of a printer; as he wrote to Mason when trying to persuade him to agree to Gray's posthumous works being printed at Strawberry Hill, "I am neither ashamed of being an author, or a bookseller. My mother's father was a timber-merchant, I have many reasons for thinking myself a worse man, and none for thinking myself better: consequently I shall never blush at doing anything he did. I print much better than I write, and love my trade."[59] And again, toward the end of the active life of the Press, when presenting Lady Ossory (1737/8–1804) with a copy of *Bishop Bonner's Ghost,* he claimed that

> Though I am [an] old printer retired from business, one cannot help now and then rubbing up one's old calling to oblige a friend; and as your Ladyship used to deal at my shop, I thought it my duty to present you with this small tribute in acknowledgement of former favours, and hope you will receive it favourably from your ancient tradesman.[60]

Having unwrapped the layers of self-deprecation from these statements, however, we should not be deluded into thinking that Walpole ever soiled his own fingers in the inky trade; the printers he employed were his workmen, and even Kirgate, after more than twenty years as printer and then secretary or amanuensis, was still referred to as a servant.[61]

Strawberry Hill Press imprints, meanwhile, continued to be collected. In addition to his immediate correspondents, there was a closer circle of collectors and occasional correspondents such as Bull (to whom Walpole wrote that "Your extreme and unmerited partiality entitle you to everything that you can think worth possessing of mine or from my press") and Storer, and then a wider and more diverse body of collectors beyond them.[62] Storer presented his library to Eton College on his death, and his Strawberry Hill books, many extra-illustrated, and typically handsomely bound in red morocco, now adorn the columned splendor of the College Library. One of his Strawberry Hill titles not at Eton is the copy of the very scarce *The Mysterious Mother* (1768), presented to him by Walpole's executrix, Anne Damer; it is in Storer's red half-morocco, but bound in before the title page is the original upper wrapper, inscribed in pen on the verso "To A.M. Storer Esqr. Devonshire Place. With Mrs. Damers Compliments"; on the recto of the wrapper, Storer has added the note "Never change the binding of this Book—A.M.S—."[63] The enthusiasm of some collectors, however, particularly print collectors such as the voracious Joseph Gulston (1744/5–1786), and the enhanced value of prints that Walpole and the early Grangerizers had provoked, created commercial pressures on print prices generally and the acquisition of Strawberry Hill items in particular that Walpole found increasingly distasteful. As he recorded in his note "Of Collectors of English Portrait Prints" in his manuscript "Book of Materials," the two principal culprits were

7
Artist unknown, *Strawberry Hill,* probably printed 1775 or later. Engraving printed in red and black, trial pull by Kirgate, 8.7 x 11.5 cm (image)

> Mr. Richard Bull and Mr. Gulston, who were indefatigable, and the former especially in little more than a year amassed a most prodigious collection. These two gentlemen spared no expense, and thence raised the prices so exceedingly, that heads which used to be sold for sixpence or less, were advanced to five shillings. They both have given five guineas for a scarce single print that two years before would not have fetched two shillings. But the greatest mischief they did was in cutting books to pieces for a single print or two.[64]

In a revealing letter of 1787 to Lady Ossory, Walpole complains about the artist and author Samuel Ireland (d. 1800), whom he accused of bribing one of his engravers to sell Ireland a copy of the print *Reason, Rectitude & Justice appearing to Christina de Pisan, and promising to assist her in writing La Cité des Dames* (Fig. 8), the frontispiece to the very scarce *Postscript to the Royal and Noble Authors* (1786, forty copies). In the same letter he objects to Gulston also bribing his engravers to obtain copies of the prints illustrating the 1784 edition of the *Description* of Strawberry Hill, and to the antiquarian Richard Gough (1735–1809) printing without his consent a list of some of those prints in his *British Topography* (1780).[65] The dual nature of the Press, by which part of its output was marketed for sale by London booksellers and so readily accessible, while part was purely intended for circulation among Walpole's immediate friends, was irresistible to collectors, then and since.

Walpole's own "List of the Books printed at Strawberry-Hill" at the end of the text of the 1784 edition of the *Description* of Strawberry Hill is the first catalogue of the productions of the Press, followed by the list given by John Pinkerton (1758–1826) in *Walpoliana* (1799). Pinkerton notes that his list was "copied in Mr. Walpole's presence," and it is interesting that it accordingly omits both 1774 and 1784 printings of the *Description* of Strawberry Hill, which Walpole kept from prying eyes.[66] Catalogues of the Press then proceed through George Baker (1810), Dibdin (in the second edition of *Bibliomania,* 1811), and the booksellers Major (1812) and McCreery (1813). Then there was John Martin's *Bibliographical Catalogue of Books Privately Printed* (1834), the second volume of which contains a far more detailed account, in the compilation of which Martin acknowledges that he was assisted by the collectors Grosvenor Charles Bedford (the son of Walpole's deputy at the Exchequer, who extra-illustrated his copy

8
Artist unknown, *"Reason, Rectitude & Justice appearing to Christina de Pisan, and promising to assist her in writing La Cité des Dames. From an illumination in the Library of the King of France,"* 1786. Engraving, 14.6 x 16.2 cm (sheet). Frontispiece to *Postscript to the Royal and Noble Authors*

of the *Description*), Sir Alexander Johnston (whose wife, Hazen notes, had inherited the books and manuscripts that Walpole bequeathed to Anne Damer), and in particular a collector from Gravesend by the name of Robert Peirce Cruden (1775–1847). Cruden is interesting as an example of an early collector with no direct family or social connection to Walpole; he is also interesting as there survive two letters from him, written in March 1819 to George Perfect Harding (1779/80–1853), the son of Silvester Harding (1745–1809), to whose roles in the dissemination of Strawberry Hill material I shall return. The letters instruct Harding to act as his agent in the sale of his books. The sale seems not to have happened, as fifteen years later Martin described Cruden's holdings as "the most curious and most perfect collection of the productions of the Strawberry Hill press," but the letters stress that the material not only contained several proof copies (in 1825 at the Baker sale, Cruden was to acquire an ultimate "proof" rarity, one of the two known copies of the short version of the 1774 *Description* of Strawberry Hill), but also, as Cruden explained to Harding, "What makes my Collection valuable, is that not only those I purchased from you—but all the others I have are *uncut*."[67]

The great Strawberry Hill sale of 1842, when the seventh Earl Waldegrave sold Walpole's collections, put many Strawberry Hill titles into circulation and generated further documentation on the imprints of the Press. Quite apart from the information to be gleaned from George Robins's auction catalogue, bookdealers who had bought deeply at the sale, such as Thomas Thorpe and William Strong of Bristol, in turn catalogued and described their purchases for resale.[68] Then in 1864 the bookseller H.G. Bohn, in revising Lowndes's *Bibliographer's Manual,* included in his Appendix volume a listing for the Strawberry Hill Press, while in 1893 Austin Dobson's influential *Horace Walpole: A Memoir* concluded with a bibliography of the books from the Press (though not the detached pieces). By now Walpole's own reputation, which had languished in the mid-nineteenth century, was beginning to revive, but without question the achievements of the Press and the continuing trade among the company of book collectors had helped to preserve it. Walpole's playful claim in 1789 to Lady Ossory that "I shall sit mighty low on the bench of authors; but Kirgate and I shall not give place to many printers in the offices of the Temple of Fame" was more prescient that he might have wished.[69]

9 (Cat. 1.1)
Probably George Perfect Harding, *"End View of Strawberry hill before the Gallery,"* ca. 1790s after a lost drawing of 1754. Pencil and watercolor, 21 x 33 cm (image), in lined mount. Laid into Kirgate's elephant folio collection of Strawberry Hill images. The Lord Waldegrave of North Hill

Plagues with Printers and Gallantries of the Press

We do not know the appearance of Walpole's first printing office. He records in the "Short Notes" of his life and in the *Journal of the Printing-Office* that the press was erected on 25 June 1757, and two letters to Lord Strafford confirm that it was to the west of the house, on the site of what became the Gallery and the Round Tower.[70] There is only one image recording (rather than hinting at) the presence of cottages or service buildings to the west of the house that Walpole acquired from Mrs. Chevenix. This is a late-eighteenth-century watercolor (Fig. 9; Cat. 1.1.), copied—probably by George Perfect Harding—from an earlier drawing (now lost, but possibly by Müntz) that Michael Snodin has dated 1754 from the state of the east front.[71] It shows, on the left, a two-storied red brick building with two pitched roofs, which Snodin speculates was the subject of the "new kitchen, Alterations to the Cottage & Work to the New Room" that in 1748 constituted Walpole's first building work at Strawberry Hill, and contained the service rooms excluded from the Chevenix house by Walpole's alterations.[72] It was the cottage that Walpole converted into a printing office, according to his letter of August 1757 to Mann; and the *Strawberry Hill Accounts* records the costs incurred in doing so: "Nov. Printing house £18-17-9 1/2" and "Bricklayers bill for printing house £7-14-0."[73]

This is also where Walpole first experienced an employer's frustrations with his printers. His original printer, the Irishman William Robinson, was engaged in June 1757 and printed the first productions of the Press: Gray's *Odes,* Hentzner's *Journey into England,* the *Catalogue of Royal and Noble Authors,* Walpole's own *Fugitive Pieces,* Lord Whitworth's *Account of Russia,* and Spence's *Parallel,* the first six items in Hazen's *Bibliography.* Although Walpole recorded two months after Robinson's arrival that "he has the most sensible look in the world; Garrick said he would give any money for four actors with such eyes," he also recounted how Robinson had left in a drawer for Walpole to find a far from sensible letter, ostensibly written to a friend in Ireland, full of flowery phrases ("I date this from shady bowers, nodding groves and amaranthine shades...") and extravagant praise of his employer, "who is very studious, and an admirer of all the liberal arts and sciences."[74] Walpole claimed to have laughed for an hour at this extravagant rhapsody, though for all its naïveté, it does in a simplistic way reflect something of Walpole's self-fashioning: the intelligent yet self-contained expression that stares out at us from Müntz's delicate wash drawing of Walpole in his library was nothing if not studious and illustrative of an admirer of the liberal arts. It is seen here in George Perfect Harding's copy (Fig. 10), a detail of the section of the Müntz image showing Walpole posed in his chair, book in hand, other books on the floor before him, his dog at his side, and behind him a Roman ossuary beneath the open window, with its view across the Thames to Richmond. In December 1758 Robinson began printing one of the most ambitious undertakings of the Press, Lucan's *Pharsalia,* with the manuscript notes of Dr. Bentley, selected and revised by his son, Walpole's designer Richard Bentley. Published in 1760, this is generally acknowledged to be the most typographically distinguished book of the Press. But three months later Walpole announced that "my press is at a stop; my printer, who was a foolish Irishman, and who took himself for a genius, and who grew angry that I thought him extremely the former, not the least of the latter, has left me, and I have not yet fixed upon another."[75]

Robinson was followed by Benjamin Williams, who came to Strawberry Hill on 29 March and "went away" on 25 May.[76] As he was about to leave, Walpole wrote to Sir David Dalrymple to inquire whether he could recommend a hardworking and sober Scots printer, as "One printer has left me, from his impertinence; another, from his imposition." Walpole explained his terms: "I give a guinea a week, not lodging nor board. This is more than even a London workman gets. He must both compose and take off, and I shall expect him to be very diligent. This is all I *engage* to pay. If he pleases me he will find the place much more advantageous."[77] Three weeks later he was able to write to Dalrymple to announce that he had found a new printer

> who comes exceedingly well recommended from Mr [Samuel] Richardson, to whose character I may trust, if one may trust to anything—yet, I own, the plague I have had with my two other printers has almost discouraged me from an amusement, which has given me as much trouble as if it was a profession: the last fellow has robbed a man and run away with his wife.[78]

Despite the strong recommendation, the new printer fared even worse than his predecessors; the *Journal of the Printing-Office* entry for June 19 records that "James Lister, a new Printer, came; staid but a week"; but on 16 July arrived Thomas Farmer, "a very industrious young man" who provided more than two years of continuity.[79]

Meanwhile Walpole, who was planning the next major building campaign to enlarge Strawberry Hill by extending it to the west, providing a Gallery terminating in a Round Tower, noted in the *Journal of the Printing-Office* for 28 August 1759 that he "began to build the new printing House." On 30 October he was able to announce to Lord Strafford that "My new printing-house is finished, in order to pull down the old one, and lay the foundations next summer of my round tower."[80] The *Journal*

10
George Perfect Harding, after J.H. Müntz, *Horace Walpole in His Library,* ca. 1790s. Pen and ink and wash, 17.8 x 12.4 cm (image)

notes that printing commenced on 1 May 1760, and on 30 June Walpole paid the bill "for the new printing House £181:13:10 1/2."[81] It was built of red brick with a tiled roof and was situated in Walpole's farmyard, to the southwest of the house (see Figs. 24–27; Catalogue 1.3–1.6).

The position of the Printing House can most clearly be seen on three later maps of the estate, two of them from 1790–91 and drawn by John Carter, and the third by George Perfect Harding and dated 1801—both artists also being illustrators of the Printing House itself.[82] The two very similar plans by Carter, unsigned but with the text in his distinctive handwriting, are bound into Walpole's and Richard Bull's extra-illustrated copies of the *Description* now at the Lewis Walpole Library (Figs. 11 and 12; Catalogue 2.2 and 2.3). Both are oriented with north to the right. To the southwest of the house is the prominent rectangular block of Walpole's new Offices (marked "D" in Walpole's copy and "C" in Bull's copy), built in 1790 to a design provided sixteen years earlier by the architect James Essex (1722–1784).[83] To the west of that (above it on the two plans), and opening onto the Teddington road, is Walpole's farmyard. Walpole's copy of the plan clearly shows the Printing House marked "F" (Bull's copy shows it marked "E"). It faced north, at right angles to the Teddington road, from which it was separated by a brick wall. To its east had stood a barn, and between them, slightly set back, Walpole in 1769 erected an ice house.[84] The barn was demolished to make way for the access required to the Offices, which also served at ground-floor level as a coach house, as seen in a gray wash drawing of ca. 1822, showing the building's rear, west facade, which faced the farmyard, the Printing House, and the Teddington road (Fig. 13).[85] In front of the Printing House is a rectangular garden plot, clearly visible

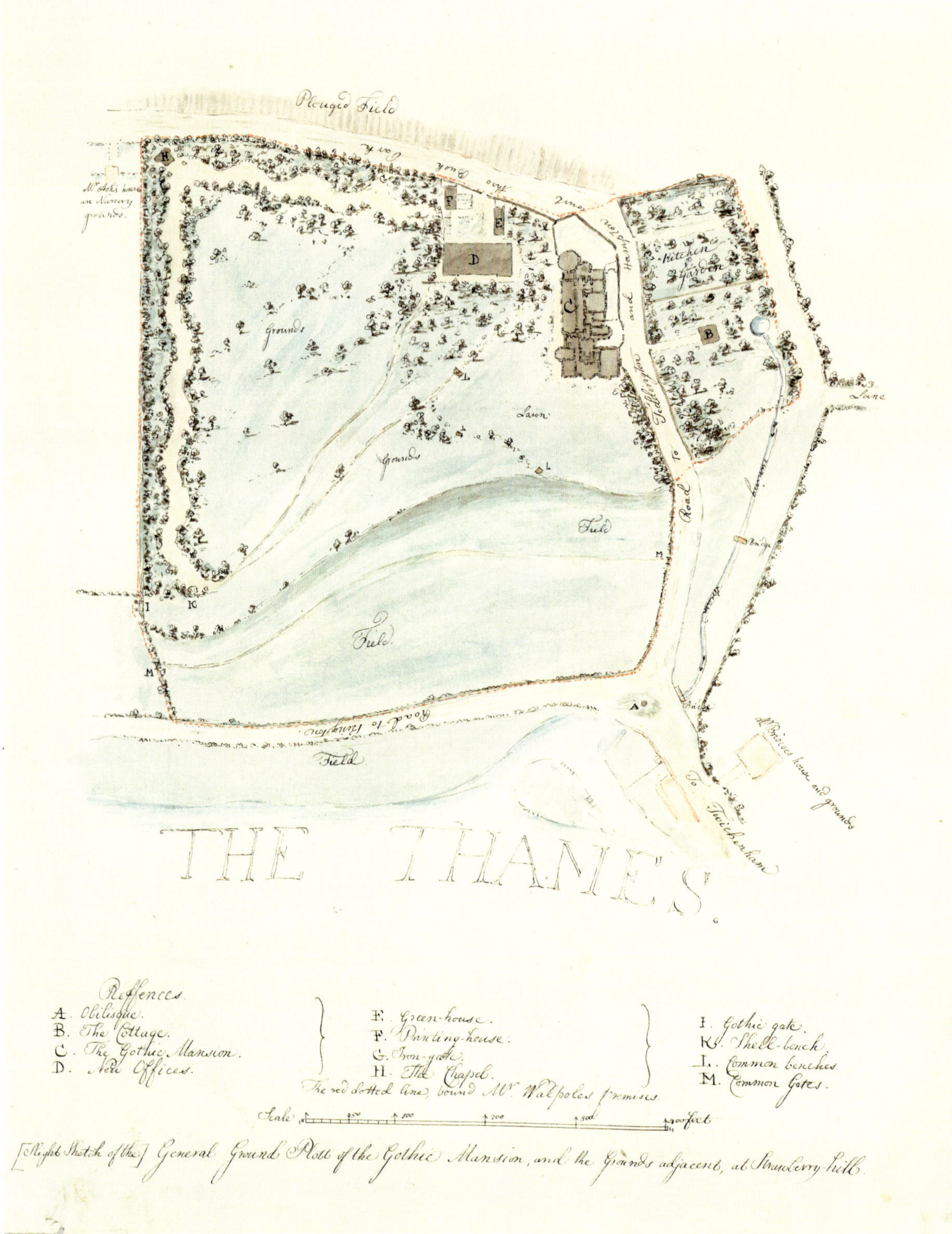
Plouged Field
Kitchen Garden
Grounds
Lawn
Grounds
Field
Field
Field
Lane
Bridge
Road to Kingston
To Twickenham
THE THAMES
Reffences.
A. Obilisque.
B. The Cottage.
C. The Gothic Mansion.
D. New Offices.
E. Green-house.
F. Printing-house.
G. Iron-gate.
H. The Chapel.
I. Gothic gate.
K. Shell-bench.
L. Common benches.
M. Common Gates.
The red dotted line, bound Mr. Walpoles premises.
Scale
[Slight Sketch of the] General Ground Plott of the Gothic Mansion, and the Grounds adjacent, at Strawberry-hill.

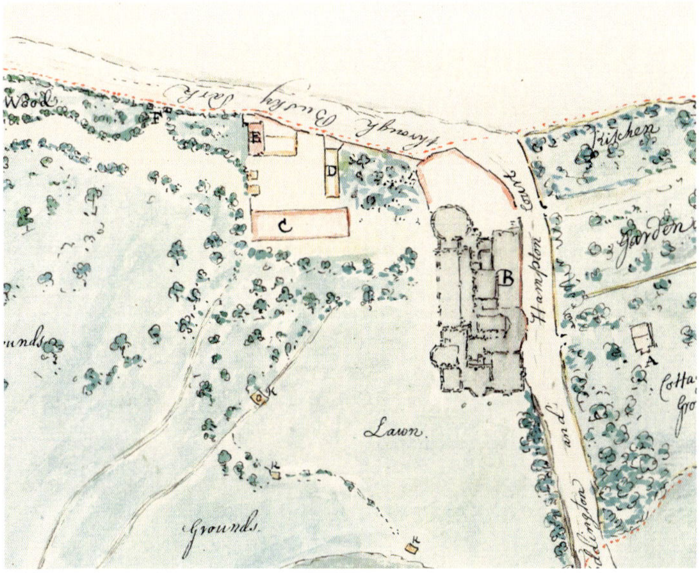

11 (Cat. 2.2) *opposite*
John Carter, *"[Slight Sketch of the] General Ground Plott of the Gothic Mansion, and the Grounds adjacent, at Strawberry-hill,"* ca. 1790–91. Watercolor with pen and ink, 46 x 31 cm. Laid into Walpole's extra-illustrated copy of the *Description* of Strawberry Hill

12 (Cat. 2.3)
John Carter, Detail of *"Slight Sketch of the Ground-Plot of Strawberry-hill,"* ca. 1790–91. Watercolor with pen and ink, 31 x 29 cm (plan, without title and key). Laid into Bull's extra-illustrated copy of the *Description* of Strawberry Hill

in both plans, and facing the Printing House across the farmyard was the Greenhouse ("E" in Walpole's copy and "D" in Bull's copy).

Harding's plan (Fig. 14; Catalogue 2.4), in Kirgate's elephant folio collection of Strawberry Hill material, is more neatly drawn, but less detailed. It is oriented so that south is at the top of the plan, which is scaled and provided with a reference key setting out the size of the various portions of the estate. Walpole's Offices and the Printing House are both clearly visible, and the ice house shown is the "new gigantic ice-house" built by Walpole in 1795, circular in plan with a porch—not on the same site as the original ice house, but set back further to the south, slightly behind the Printing House.[86]

The Printing House functioned both as a working printing office and as a diversion for distinguished visitors on the tour of Strawberry Hill, when it could be used to produce tributes of social gallantry. An early example of the social use of the Press is provided in Walpole's account to George Montagu of a visiting party of French and English aristocrats in May 1763:

> We breakfasted in the great parlour, and I had filled the hall and large cloister by turns, with French horns and clarinets. As the French ladies had never seen a printing-house, I carried them into mine; they found something ready set, and desiring to see what it was, it proved as follows;
>
> The Press Speaks.
> *For Madame de Boufflers.*
> The graceful fair, who loves to know,
> Nor dreads the North's inclement snow;

13
Artist unknown, *"The Stables at Strawberry Hill"* (rear/west facade of Walpole's Offices), early 1820s. Pen and ink and wash, 17.8 x 25 cm (image). Bound into John D. Fry's extra-illustrated copy of the *Description* of Strawberry Hill

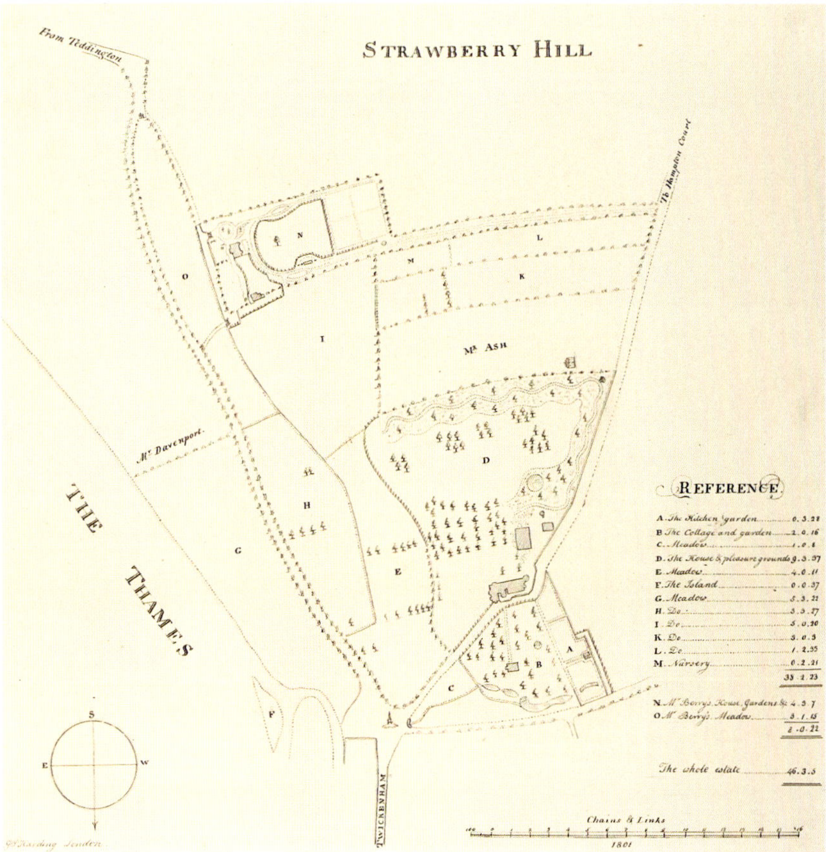

14 (Cat. 2.4)
George Perfect Harding, *"Strawberry Hill,"* 1801. Pen and ink, 27.7 x 27.7 cm. Laid into Kirgate's elephant folio collection of Strawberry Hill images. The Lord Waldegrave of North Hill

Who bids her polish'd accent wear
The British diction's harsher air;
Shall read her praise in ev'ry clime,
Where types can speak, or poets rhyme.

For Madame Dusson.
Feign not an ignorance of what I speak;
You could not miss my meaning, were it Greek.
'Tis the same language Belgium utter'd first,
The same which from admiring Gallia burst.
True sentiment a like expression pours;
Each country says the same to eyes like yours.

> You will comprehend that the first speaks English, and that the second does not; that the second is handsome, and the first not; and that the second was born in Holland. This little *gentillesse* pleased, and atoned for the popery of my house, which was not serious enough for Madame Boufflers, who is Montmorency, and *du sang du premier Chrétien;* and too serious for Madame Dusson, who is a Dutch Calvinist.[87]

There was a similar event the following summer, when after a morning shower brightened the grass and flowers and a later thunderstorm "gave a dignity of colouring to the heavens," dinner was accompanied by French horns and clarinets in the cloister; and the guests, English and French, were treated to a syllabub (milked from cows brought onto the brow of the terrace before the house) and then escorted to the Printing House, where they saw a new fashionable French song printed.[88] But the most celebrated event of this kind occurred in May 1769, as recounted in another letter from Walpole to George Montagu:

> Strawberry has been in great glory—I have given a *festino* there that will almost mortgage it. Last Tuesday all France dined there. Monsieur and Madame du Châtelet, the Duc de Liancour, three more French ladies whose names you will find in the enclosed paper, eight other Frenchmen, the Spanish and Portuguese ministers, the Holdernesses, Fitzroys, in short we were four and twenty. They arrived at two. At the gates of the castle I received them dressed in the cravat of Gibbins's [*sic*] carving, and a pair of gloves embroidered up to the elbows that had belonged to James I. The French servants stared and firmly believed this was the dress of English country gentlemen. After taking a survey of the apartments, we went to the printing-house where I had prepared the enclosed verses [Fig. 15], with translations by Monsieur de Lisle, one of the company. The moment they were printed off, I gave a private signal and French horns and clarinets accompanied the compliment. We then went to see Pope's grotto and garden, and returned to a magnificent dinner in the refectory. In the evening we walked, had tea, coffee and lemonade in the gallery, which was illuminated with a thousand, or thirty candles, I forget which, and played at whisk and loo till midnight. Then there was a cold supper, and at one the company returned to town saluted by fifty nightingales, who as tenants of the manor came to do honour to their lord.[89]

The first thing to note in relation to these escorted visits to the Printing House is that the visitors were largely French: what is being enacted here is a form of Watteauesque *fête champêtre,* in which the Press becomes part of the enchantment. On all three occasions there are French courtiers and music; in the last there is the playful use of costume (both the Grinling Gibbons carved cravat and the gloves of King James I have survived and were included in the exhibition *Horace Walpole's Strawberry Hill* at the Yale Center for British Art in New Haven and the Victoria and Albert Museum in London in 2009–10), and also the theatricality of Walpole's sense of timing in arranging for a fanfare from his band of French horns and clarinets as the complimentary verses to the French ladies were printed. On the second occasion there is the rural idyll of cows being milked before the house to provide syllabub for the visitors' delight, and Walpole's sensitivity to lighting (both natural and artificial) in the presentation of the house is apparent from the last two events. The Press is here not reaching out to a wider public, but reaching back into Walpole's historical imagination, assisting in the re-creation of a seventeenth-century courtly scene. The real world, however, could not forever be excluded; among Kirgate's collection of "Loose Pieces Printed at Strawberry Hill" is a note in his hand recording the identity of the four French ladies who received Walpole's printed verse compliments in the *festino* of May 1769, and adding that one of them, Madame du Châtelet, "was guillotined at Paris under the tyranny of that Monster Robespierre, as her husband had been before."[90]

Walpole's creative use of "a gallantry from the press" (the phrase is George Montagu's) extended throughout its working life and provided a number of the detached pieces under the heading "The Press speaks."[91] He took the trope of the press as the vehicle for revealing uncensored truth and manipulated it to present flattery as if it were discovered truth. As early as August 1757, two months after the Press opened, it celebrated the wit of Lady Townshend and the beauty of Lady Rochford, the former Miss Young, doing so with a carefully planned coup de théatre:

> I must give you some account of *les amusements des eaux de Straberri.* T'other day my Lady Rochford, Lady Townshend, Miss Bland and the new Knight of the Garter dined here, and were carried into the printing-office, and were to see the man [Robinson] print. There were some lines ready placed, which he took off; I gave them to my Lady Townshend; here they are;
>
> The Press speaks;
> From me wits and poets their glory obtain;
> Without me their wit and their verses were vain.
> Stop, Townshend, and let me but *print* what you say;
> You, the fame I on others bestow, will repay.
>
> They then asked, as I foresaw, to see the man compose; I gave him four lines out of *The Fair Penitent,* which he set, but while he went to place them in the press, I made them look

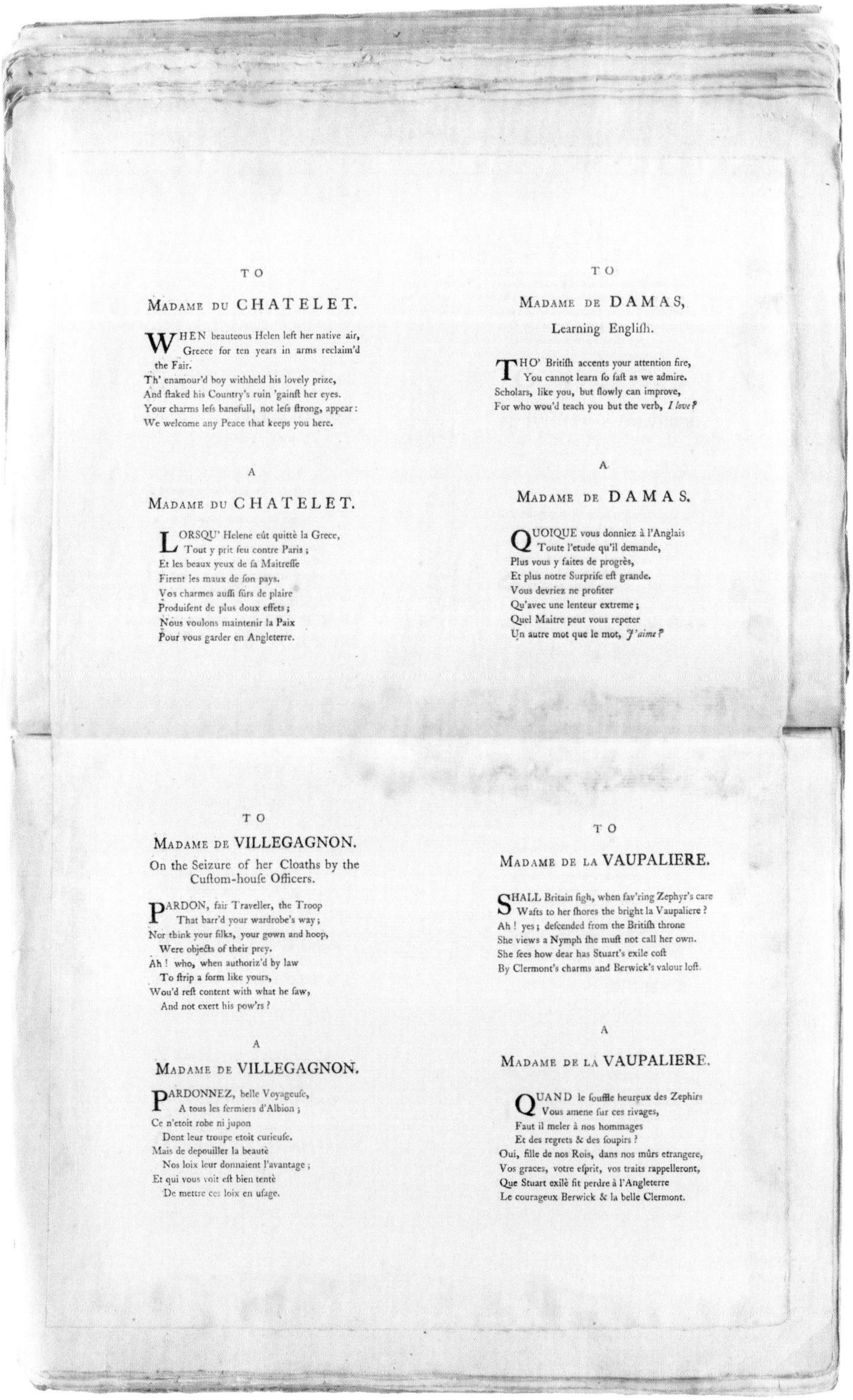

TO

MADAME DU CHATELET.

WHEN beauteous Helen left her native air,
Greece for ten years in arms reclaim'd the Fair.
Th' enamour'd boy withheld his lovely prize,
And ſtaked his Country's ruin 'gainſt her eyes.
Your charms leſs banefull, not leſs ſtrong, appear:
We welcome any Peace that keeps you here.

A

MADAME DU CHATELET.

LORSQU' Helene eût quittè la Grece,
Tout y prit feu contre Paris;
Et les beaux yeux de ſa Maitreſſe
Firent les maux de ſon pays.
Vos charmes auſſi ſûrs de plaire
Produiſent de plus doux effets;
Nous voulons maintenir la Paix
Pour vous garder en Angleterre.

TO

MADAME DE DAMAS,

Learning Engliſh.

THO' Britiſh accents your attention fire,
You cannot learn ſo faſt as we admire.
Scholars, like you, but ſlowly can improve,
For who wou'd teach you but the verb, *I love*?

A

MADAME DE DAMAS.

QUOIQUE vous donniez à l'Anglais
Toute l'etude qu'il demande,
Plus vous y faites de progrès,
Et plus notre Surpriſe eſt grande.
Vous devriez ne profiter
Qu'avec une lenteur extreme;
Quel Maitre peut vous repeter
Un autre mot que le mot, *J'aime*?

TO

MADAME DE VILLEGAGNON.

On the Seizure of her Cloaths by the Cuſtom-houſe Officers.

PARDON, fair Traveller, the Troop
That barr'd your wardrobe's way;
Nor think your ſilks, your gown and hoop,
Were objects of their prey.
Ah! who, when authoriz'd by law
To ſtrip a form like yours,
Wou'd reſt content with what he ſaw,
And not exert his pow'rs?

A

MADAME DE VILLEGAGNON.

PARDONNEZ, belle Voyageuſe,
A tous les fermiers d'Albion;
Ce n'etoit robe ni jupon
Dont leur troupe etoit curieuſe.
Mais de depouiller la beautè
Nos loix leur donnaient l'avantage;
Et qui vous voit eſt bien tentè
De mettre ces loix en uſage.

TO

MADAME DE LA VAUPALIERE.

SHALL Britain ſigh, when fav'ring Zephyr's care
Wafts to her ſhores the bright la Vaupaliere?
Ah! yes; deſcended from the Britiſh throne
She views a Nymph ſhe muſt not call her own.
She ſees how dear has Stuart's exile coſt
By Clermont's charms and Berwick's valour loſt.

A

MADAME DE LA VAUPALIERE.

QUAND le ſouffle heureux des Zephirs
Vous amene ſur ces rivages,
Faut il meler à nos hommages
Et des regrets & des ſoupirs?
Oui, fille de nos Rois, dans nos mûrs etrangere,
Vos graces, votre eſprit, vos traits rappelleront,
Que Stuart exilè fit perdre à l'Angleterre
Le courageux Berwick & la belle Clermont.

15
Verses to Madame du Châtelet, Madame de Damas, Madame de Villegagnon, and Madame de Vaupalière. Strawberry Hill Press, 1769. Letterpress, 40 x 25.3 cm (irreg.)

at something else; without their observing, and in an instant he whipped away what he had just set, and to their great surprise when they expected to see *Were ye, ye Fair,* he presented to my Lady Rochford the following lines;

> The Press speaks;
> In vain from your properest name you have flown,
> And exchanged lovely Cupid's for Hymen's dull throne;
> By my art shall your beauties be constantly sung,
> And in spite of yourself you shall ever be Young.

You may imagine, whatever the poetry was, that the gallantry of it succeeded.[92]

Over thirty years later, Walpole wrote to Lady Ossory that when he had arranged for Mary Berry and her sister Agnes to visit the Press, "I recollected my gallantry of former days, and they found these stanzas ready set."[93] The verses, praising Mary as classicist and Agnes as artist, and lamenting the inability of the Press to add to their fame, are entirely consistent with this use of the Press throughout its lifetime, celebrating Walpole's friendships and the formalities of courtesy.

Such social exchanges did not, however, entirely exclude the harsh realities of commerce and employment. The "very industrious young man" Thomas Farmer ran away at the beginning of December 1761. As Walpole lamented to George Montagu, "Just when I thought my book finished, my printer ran away, and had left eighteen sheets in the middle of the book untouched, having amused me with sending proofs. He had got into debt, and two girls with child—being two, he could not marry both Hannahs. You see my luck; I had been kind to this fellow."[94] The book was the *Anecdotes of Painting,* and Walpole also noted how Farmer's failure to print the whole number of sheets "accounts too for the difference of the ink in various sheets, and for some other blemishes"—failings that can be very visible in some copies of the second volume of the *Anecdotes of Painting,* where over-inking can cause the print to offset onto the facing page, and also (particularly if the imposition was not sufficiently exact) to show through from the page behind (Figs. 16 and 17).[95] A fifth printer, William Pratt, was immediately engaged to finish the work, and the first two volumes of the *Anecdotes of Painting* were published the following February. In May, Walpole noted that Pratt "came to live with me." We know from Walpole's letter to Dalrymple of 22 May 1759 and from his accounts kept with the *Journal of the Printing-Office* that he had previously paid his printers a guinea a week without lodging or board, but the accounts also show that in 1763 and 1764, Pratt was paid half a guinea, with Walpole paying for the accommodation for him and his wife with a Robert Taylor at 1s. 6d. a week; it appears that Pratt lodged near rather than at Strawberry Hill.[96] Pratt was to print the third volume of the *Anecdotes of Painting* and the *Catalogue of Engravers,* but as soon as Walpole began in September 1763 to print the curiosity of *The Life of*

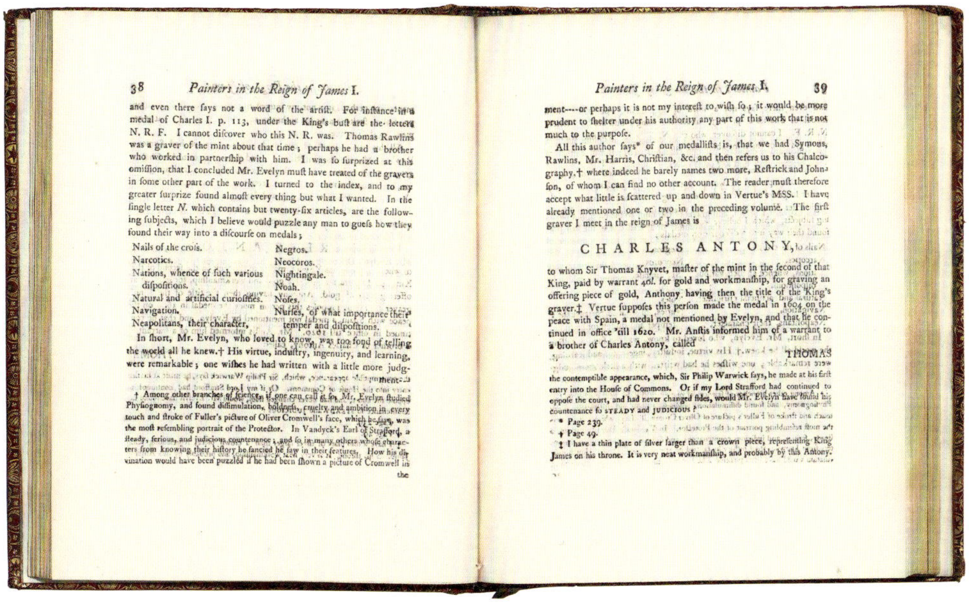

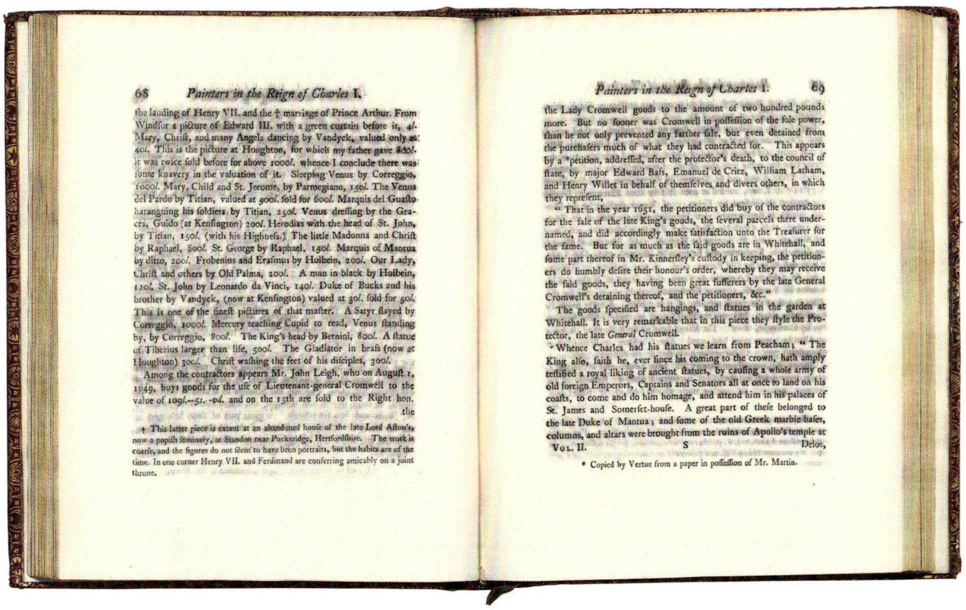

16 and 17

Horace Walpole, *Anecdotes of Painting in England.* Strawberry Hill Press, 1762. Vol. 2, pp. 38–39 and 68–69, showing offsetting caused by over-inking (fig. 16) and show-through caused by poor imposition and over-inking (fig. 17). Private collection

Edward Lord Herbert of Cherbury, he suffered fresh problems. Walpole explained the now familiar sequence of events to George Selwyn:

> I am prevented waiting on you as I intended on Sunday. There is no end of my plagues with printers. My present one, like his predecessor [Farmer], has run into debt here, and been forced to leave me and I cannot stir, as I must have my bookseller come from London and examine my printing-house, and see whether he has taken any copies away; and I must get another printer, having just begun a very curious book, that has been lent me, and which I am very much pressed to get finished—in short, the state is not more out of order than my *imprimerie.* This is very disagreeable on many accounts; on none more, than by interfering with the agreeable tour I had proposed. I don't know whether I am robbed, nor how I may have suffered by this fellow and a drunken wife, who has occasioned his extravagance.[97]

The bookseller to whom he refers is William Bathoe (d. 1768), who advised Walpole in relation to the Press and assisted with some of the accounts.[98] But it appears that Walpole's need of a printer was greater than his dissatisfaction with the improvident Pratt and his wife, as Walpole settled the bill and seems to have allowed Pratt to work off the debt, finally discharging him in December 1764.[99]

The two entries in the *Journal of the Printing-Office* for 1765, and the first entry for 1768, are worth reproducing for their significance for the future of the Press:

> 1765.
> March 18. took T. Kirgate, a new Printer, & continued reprinting of the Anecdotes.
> Aug. 24. finished everything, & discharged Printer, as I was going abroad.
>
> 1768.
> April 6th. took Kirgate the Printer again.[100]

Thomas Kirgate served Walpole as his printer, and latterly as his secretary, for virtually thirty years; effectively, he *is* the printer of the Strawberry Hill Press, even though some of the Press's more famous productions preceded him.[101] We have no portraits of his predecessors or images of them at or about the Press, but we do of Kirgate (see section 3 of the Iconography, which catalogues the surviving images of Kirgate), as well as some correspondence, and even the sale catalogue of his not inconsiderable library and print collection. We also have the enticing duality of Kirgate as faithful, long-standing employee, shabbily treated on Walpole's death, and Kirgate as forger.

When Kirgate was reengaged in 1768, the Press had been silent for nearly three years. His first work was Président Hénault's tragedy of *Cornélie* (1768), two hundred copies, followed by the fifty copies of Walpole's own tragedy *The Mysterious Mother;* the *Poems* of Mr. Hoyland (1769), printed for his benefit; and Walpole's abortive quarto edition of his own works. From then until 1774, Kirgate printed the last volume of the *Anecdotes of Painting,* the *Mémoires de Grammont,* the *Seven Original Letters from Edward VI,* the two numbers of the *Miscellaneous Antiquities,* and the first edition of the *Description of the Villa of Horace Walpole…at Strawberry Hill.* This was his most sustained period of activity as printer. Walpole noted in the *Journal of the Printing-Office* that when the *Description* was finished in 1774, "The Printer was employed in making Catalogues of my Portrait-prints and other ways," of which the principal was as secretary during Walpole's recurrent attacks of gout.[102] One of the first instances of this is a letter to Lord Hardwicke (1720–1790) in Kirgate's hand, conjecturally dated by the editors of Walpole's correspondence to January 1773, and in December that year Walpole acknowledged that "my printer is turned into a secretary."[103] This meant that over the last twenty years of Walpole's life, the balance of Kirgate's responsibilities slowly shifted away from the Press. As the severity of Walpole's gout increased, he became increasingly dependent on Kirgate's services; as he playfully announced to Lady Ossory in 1782, "I am in debt for some others [letters], but my Secretary of State must answer them, for I find our royal breast is as tired as our hand."[104] The ostensible similarity of Walpole's and Kirgate's hands has long caused confusion, though Kirgate's writing tends to be slightly more angular than his employer's.

In addition to his role as amanuensis, we see Kirgate involved in a whole range of activities: being sent to London or Richmond on an errand, attending the sale of Sir Joshua Reynolds's pictures, checking references, framing a watercolor by Agnes Berry, meeting the coroner after a servant hung himself on the grounds of the estate, being put forward to take a disordered ward of Walpole's into Kent, showing the house to visitors, and even copying Walpole's plan of the picture hang in the Long Gallery "as a Guide for the Housekeeper."[105] But these duties were combined with the later productions of the Press, of which the most substantial item was the expanded 1784 edition of the *Description of the Villa of Mr. Horace Walpole…at Strawberry Hill,* which with its plates and appendices was not completed for another seven years.[106] There was also the French translation by the Duc de Nivernois of Walpole's *Essay on Modern Gardening* and the great rarity of his *Hieroglyphic Tales* (both 1785). When in 1789 Kirgate printed More's *Bishop Bonner's Ghost,* Walpole had to explain to the author that the press, which in Walpole's last years was used for little more than very occasional verses and labels, "has lain fallow for some time, and my printer must prepare ink-balls [for inking the type] etc."; eight days later he lamented that there would be a further delay, as his rolling press was out of order, and he had to send the whole impression to London to have the copperplate vignette of Strawberry Hill on the title page printed.[107]

There is nothing in Walpole's lifetime to suggest that his relations with Kirgate were other than good. In 1791 he mentioned Kirgate's kindness in the case of the servant who committed suicide, and he noted without complaint that his secretary "is grown deafish."[108] In 1786 he had presented Kirgate with a drawing of "A Modern

Venus" by Miss Hoare of Bath for him to have it engraved for his own benefit.[109] He appears from about 1790 to have permitted Kirgate to undertake a certain amount of printing on his own account—occasional verses, notices, and cards of address.[110] Yet on his death in 1797, Walpole left Kirgate only £100, in contrast to the £1,500 (with an annuity of £25 a year) for his valet, Philip Colomb, and the same amount to his Clerk at the Exchequer, and consequently attracted much opprobrium then and since.[111] Joseph Farington the diarist noted that the will lacked "due sense of services rendered him."[112] John Pinkerton darkly commented that Kirgate's "modest merit was supplanted in Lord Orford's will by intriguing impudence, as always happens," and even Mary Berry acknowledged of Kirgate that "whether from carelessness, or from an inadequate idea of his necessities, it seems to be a melancholy fact that he was not sufficiently provided for by will."[113]

In 1931 W.S. Lewis jumped to Walpole's defense in *The Forlorn Printer,* showing how the number of Strawberry Hill imprints in the sale of Kirgate's library and collection after his death in 1810 strongly suggested that Kirgate had been secreting away additional copies.[114] Mrs. Toynbee, the editor of the edition of Walpole's correspondence that preceded Lewis's, had already noted in 1903 that Kirgate, who made transcriptions for Walpole of his letters to Sir Horace Mann, had kept copies of the passages suppressed in the transcriptions; these were handed over by Kirgate's daughter Eleanor Thomas in 1810 for destruction, her surviving memorandum stating that the extracts "were in the Possession of my Father the late Mr. Thomas Kirgate at his Death" and acknowledging that they "were not intended by his Lordship to be either transcribed or printed."[115] Then in 1942 Hazen's *Bibliography* of the Press confirmed beyond all reasonable doubt that Kirgate had also produced at the Press a series of copies of Strawberry Hill printings—effectively forgeries—that could be marketed to provide him with the pension that Walpole had failed to supply.[116] A long debate about the relative status of ordinary- and thick-paper copies of Gray's *Odes,* for example, was at last resolved when Hazen established that the thick-paper copies were Kirgate's completely separate and much later printing. He initially, from study of Kirgate's reprints of Gray's *Odes* and the Countess Temple's *Poems,* considered that they dated from 1790 or later for the *Odes,* and probably 1794 or later for the *Poems.* But by the time his *Bibliography* was reprinted in 1973, Hazen had concluded that it was safe to assume that all or most of the reprints, which include a number of the detached pieces, dated from 1797, the year of Walpole's death.[117] As an example, the "Verses sent to Lady Charles Spencer" (1764) was reprinted by Kirgate, who replaced the original uppercase "P" of "Occasioned by saying she was low in Pocket" in the heading with a lowercase "p" (Figs. 18 and 19). Barker has recently made the point that it may be misleading to impose modern bibliographical preoccupations with authenticity on Kirgate's conduct, using the analogy of Grangerizers and print collectors cheerfully accepting modern copies of portraits when the rare original was unobtainable.[118] There is, however, no evidence to suggest that Kirgate, in his dealings in Strawberry Hill material, ever acknowledged the status of his reprints, and given the financial benefits involved, it is hard not to read silence as deception.

With these issues in mind, Kirgate's sale catalogue (Fig. 20) is a revealing document. Kirgate died on 11 June 1810, and the sale of his collection was conducted by King and Lochée of King Street, Covent Garden, taking place over eleven days commencing on 3 December that year. The first two days were devoted to books, including eighty-three lots of the "Earl of Orford's Works, chiefly printed at Strawberry-Hill," followed by seven days of prints, one day of glass and coins, and as a Supplement, a final day of catalogues of books and prints (which look like a dealer's reference library), more prints, some copperplates, and another eighty-nine lots of "Books Printed chiefly at Strawberry-Hill." The total of 172 Strawberry Hill-related lots is an extraordinary assemblage, even allowing for the possibility of Kirgate having retained some of the additional copies or "overs" traditionally produced by printers when printing a

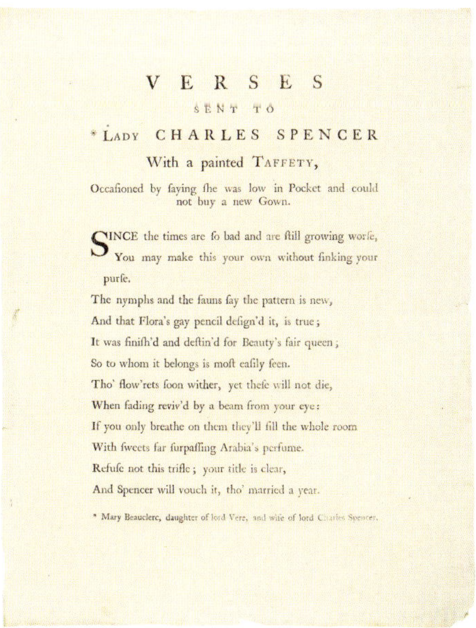

VERSES
SENT TO
*Lady CHARLES SPENCER
With a painted Taffety,
Occaſioned by ſaying ſhe was low in Pocket and could not buy a new Gown.

Since the times are ſo bad and are ſtill growing worſe,
You may make this your own without ſinking your purſe.
The nymphs and the fauns ſay the pattern is new,
And that Flora's gay pencil deſign'd it, is true;
It was finiſh'd and deſtin'd for Beauty's fair queen;
So to whom it belongs is moſt eaſily ſeen.
Tho' flow'rets ſoon wither, yet theſe will not die,
When fading reviv'd by a beam from your eye:
If you only breathe on them they'll fill the whole room
With ſweets far ſurpaſſing Arabia's perfume.
Refuſe not this trifle; your title is clear,
And Spencer will vouch it, tho' married a year.

* Mary Beauclerc, daughter of lord Vere, and wife of lord Charles Spencer.

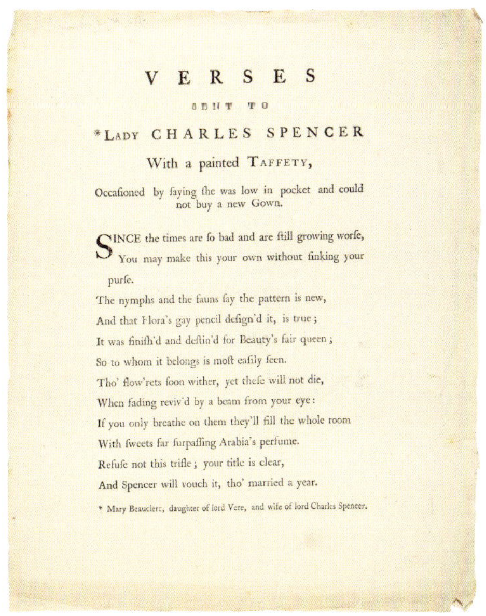

VERSES
SENT TO
*Lady CHARLES SPENCER
With a painted Taffety,
Occaſioned by ſaying ſhe was low in pocket and could not buy a new Gown.

Since the times are ſo bad and are ſtill growing worſe,
You may make this your own without ſinking your purſe.
The nymphs and the fauns ſay the pattern is new,
And that Flora's gay pencil deſign'd it, is true;
It was finiſh'd and deſtin'd for Beauty's fair queen;
So to whom it belongs is moſt eaſily ſeen.
Tho' flow'rets ſoon wither, yet theſe will not die,
When fading reviv'd by a beam from your eye:
If you only breathe on them they'll fill the whole room
With ſweets far ſurpaſſing Arabia's perfume.
Refuſe not this trifle; your title is clear,
And Spencer will vouch it, tho' married a year.

* Mary Beauclerc, daughter of lord Vere, and wife of lord Charles Spencer.

18
Verses sent to Lady Charles Spencer With a painted Taffety. Strawberry Hill Press, 1764. Letterpress, 27.5 x 21.5 cm (irreg.)

19
Verses sent to Lady Charles Spencer With a painted Taffety. Kirgate reprint, ca. 1797. Letterpress, 27.2 x 21.5 cm (irreg.)

limited edition. It consists of nine copies of *Grammont* (of only one hundred printed), three copies of the *Anecdotes of Painting* (of which we have noted that Walpole complained that he had not a spare copy left), four (of forty) copies of the *Postscript to the Royal and Noble Authors,* thirteen (of two hundred) of the *Letter to the Editor of the Miscellanies of Thomas Chatterton* (1779), one (of fifty) of the *Mysterious Mother,* and sixteen (of two hundred not sent to Paris) of the translation of the *Essay on Modern Gardening.* Lot 380 on the second day was a copy of the *Hieroglyphic Tales,* of which the "Short Notes of the Life of Horatio Walpole" says there were five copies, the *Journal of the Printing-Office* says there were six, and Kirgate's sale catalogue says there were seven; one suspects that Kirgate knew best.

There were, not surprisingly, multiple copies of the titles that Kirgate is known to have reprinted—on the last day, eleven copies of Gray's *Odes* and ten copies of the Countess Temple's *Poems.* Two of the copies of *Grammont* were "Avec Notes par Horace Walpole, uncut"; one copy of the *Anecdotes of Painting* had notes by Walpole and Kirgate; and one copy of the *Essay on Modern Gardening* had two letters from Walpole to the Duc de Nivernois, and many prints. A highlight of the sale was "A COMPLETE SET of the loose Pieces printed at Strawberry Hill, upwards of Sixty in number, the only complete set extant, many of which are extremely rare, if not unique, and MS. Notes by T. Kirgate"; but there were in addition to that eight lots of Loose Pieces, three of them with more than thirty items each. And a copy of the sale catalogue at the Lewis Walpole Library confirms that one lot was withdrawn—Kirgate's transcriptions of Walpole's letters to Sir Horace Mann.[119]

But Kirgate's continuing involvement with Strawberry Hill was not limited to the productions of the Press. The 546 lots of prints included, as lots 817 to 843 (Fig. 21), more than two hundred drawings and prints of the house and its collections by artists including Carter, J.C. Barrow, Edwards, Harding, and Kirgate himself, ranging from seven drawings of Strawberry Hill in its original state and in 1753, and four drawings by Carter of ancient busts, to "Six Internal Views of Strawberry-Hill, *printed on sattin.*" These are in fairness part of a substantial and varied body of prints that was part stock and part collection, but they would appear to be ideal material for any aspiring Grangerizer.

It is worth remembering that Walpole's will was dated May 1793, when Kirgate's reprinting and dealing activities lay in the future, and it is not known to what extent, if at all, Walpole had suspicions of Kirgate's conduct when he determined his legacy. There are, however, two references in Walpole's correspondence that may imply concerns about Kirgate's honesty. In 1780 Walpole explained his refusal to take up the suggestion of the Prince de Beauvau that he print portraits and characters from the papers of Mme. du Deffand, which he had received after her death in 1780, on the grounds that "as it is impossible for me to be secure that my printer would not reserve

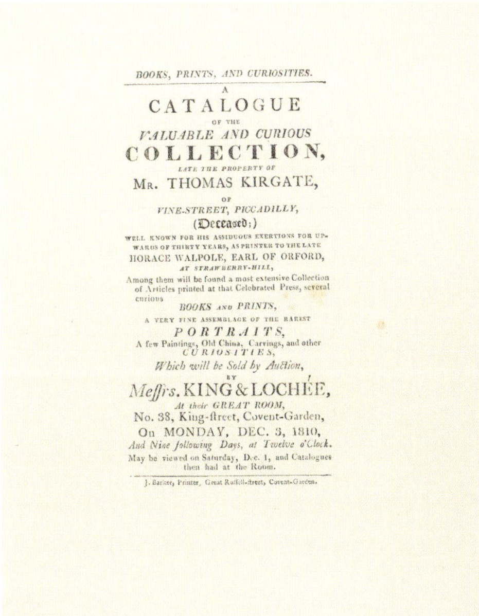

BOOKS, PRINTS, AND CURIOSITIES.

A CATALOGUE OF THE VALUABLE AND CURIOUS COLLECTION, LATE THE PROPERTY OF MR. THOMAS KIRGATE, OF VINE-STREET, PICCADILLY, (Deceased;) WELL KNOWN FOR HIS ASSIDUOUS EXERTIONS FOR UPWARDS OF THIRTY YEARS, AS PRINTER TO THE LATE HORACE WALPOLE, EARL OF ORFORD, AT STRAWBERRY-HILL, Among them will be found a most extensive Collection of Articles printed at that Celebrated Press, several curious BOOKS AND PRINTS, A VERY FINE ASSEMBLAGE OF THE RAREST PORTRAITS, A few Paintings, Old China, Carvings, and other CURIOSITIES, Which will be Sold by Auction, BY Meſſrs. KING & LOCHÉE, At their GREAT ROOM, No. 38, King-ſtreet, Covent-Garden, On MONDAY, DEC. 3, 1810, And Nine following Days, at Twelve o'Clock. May be viewed on Saturday, Dec. 1, and Catalogues then had at the Room.

J. Barker, Printer, Great Ruſſell-ſtreet, Covent-Garden.

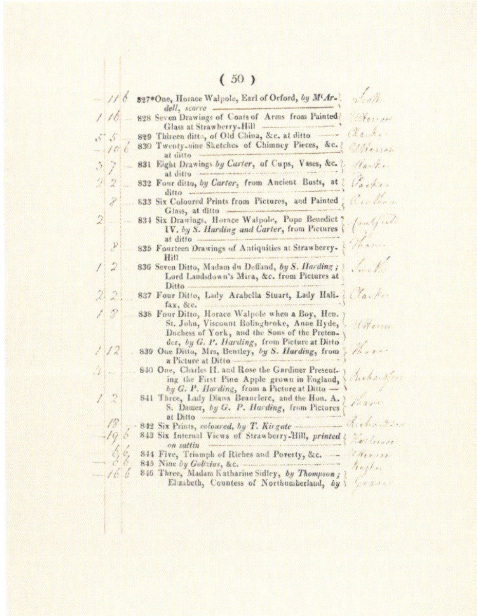

(50)

827*One, Horace Walpole, Earl of Orford, *by McArdell, scarce*

828 Seven Drawings of Coats of Arms from Painted Glass at Strawberry-Hill

829 Thirteen ditto, of Old China, &c. at ditto

830 Twenty-nine Sketches of Chimney Pieces, &c. at ditto

831 Eight Drawings *by Carter*, of Cups, Vases, &c. at ditto

832 Four ditto, *by Carter*, from Ancient Busts, at ditto

833 Six Coloured Prints from Pictures, and Painted Glass, at ditto

834 Six Drawings, Horace Walpole, Pope Benedict IV. *by S. Harding and Carter*, from Pictures at ditto

835 Fourteen Drawings of Antiquities at Strawberry-Hill

836 Seven Ditto, Madam du Deffand, *by S. Harding*; Lord Landsdown's Mira, &c. from Pictures at Ditto

837 Four Ditto, Lady Arabella Stuart, Lady Halifax, &c.

838 Four Ditto, Horace Walpole when a Boy, Hen. St. John, Viscount Bolingbroke, Anne Hyde, Duchess of York, and the Sons of the Pretender, *by G. P. Harding*, from Picture at Ditto

839 One Ditto, Mrs, Bentley, *by S. Harding*, from a Picture at Ditto

840 One, Charles II. and Rose the Gardiner Presenting the First Pine Apple grown in England, *by G. P. Harding*, from a Picture at Ditto

841 Three, Lady Diana Beauclerc, and the Hon. A. S. Damer, *by G. P. Harding*, from Pictures at Ditto

842 Six Prints, *coloured, by T. Kirgate*

843 Six Internal Views of Strawberry-Hill, *printed on sattin*

844 Five, Triumph of Riches and Poverty, &c.

845 Nine *by Goltzius*, &c.

846 Three, Madam Katharine Sidley, *by Thompson*; Elizabeth, Countess of Northumberland, *by*

20 and 21

Title page and page 50, *A Catalogue of the Valuable and Curious Collection, Late the Property of Mr. Thomas Kirgate, of Vine-Street, Piccadilly (Deceased;).* London, 1810

a copy or copies for himself, it must risk what the prince himself apprehended, their being published; and that nothing could induce me to run a hazard of that kind." Three years later, on hearing a false rumor that the Printing House had been robbed, William Mason (not perhaps, in fairness, the most unimpeachable witness) wrote to Walpole that "I hope you have lost nothing of great consequence. On this topic I think you might employ another hand."[120] It may well be that Walpole knew or suspected something of Kirgate's hoarded copies, but there is also a sense that outside the immediate range of his family and social circle, Walpole was not a naturally generous man.[121]

Kirgate, meanwhile, responded by printing his poignant adieu to his own press, "The Printer's Farewell to Strawberry-Hill" where, "neglected and forgot," he bemoaned the cruel fate by which he was forced to shut his Printing House door behind him and face the bustle of life after years of service (Fig. 22). The poem was written for him by his friend the artist and bookseller Silvester Harding, whom he had met in the early 1780s when Harding was introduced to Strawberry Hill by Walpole's correspondent Nathaniel Hillier (ca. 1707–1783).[122] A copy of the poem in Kirgate's own hand is bound into James Hughes Anderdon's compilation of portrait prints and biographical notes, the 105 volumes of the "Collectanea Biographica" in the British Museum. The poem is signed with Kirgate's initials and dated October 1797,

THE PRINTER's FAREWELL

TO STRAWBERRY-HILL.

ADIEU! ye Groves and Gothic Tow'rs,
Where I have ſpent my youthful Hours,
Alas! I find in vain:
Since he who could my Age protect,
By ſome myſterious, ſad neglect,
Has left me to complain!

For thirty Years of Labour paſt,
To meet ſuch ſlight Reward at laſt,
Has added to my Cares:
To quit the quiet Scenes of Life,
T'encounter Bus'neſs, Buſtle, Strife,
Hangs heavy on my Years.

Farewell! my PRINTING-HOUSE, farewell!
Where I no more ſhall calmly dwell,
Within thy peaceful Door:
No more in Converſation free,
Enjoy my Friend and ſip my Tea;
Ah! no; thoſe Days are o'er.

On thee, my Fellow-Lab'rour, dear,
My PRESS, I drop the ſilent Tear
Of Pity, for thy Lot;
For thou, like me, by Time art worn,
Like me, too, thou art left forlorn,
Neglected and forgot!

October, 1797. T. K.

22

The Printer's Farewell to Strawberry-Hill. Strawberry Hill Press, 1797.
Letterpress, 27.3 x 21.5 cm (irreg.)

and the references to the "Printing House" and "Press" appear all the more personal for being underlined twice, consistent with the printing convention by which double underlining in manuscript is translated into small capitals in the printed text.[123] For Kirgate, "The Printer's Farewell" was his melancholy valediction, the product of the final parting of printer and press, and he annotated the copy of the poem here illustrated: "The Last Thing Printed at Strawberry hill."[124] In practice, though, it appears from the researches of Hazen that if it was indeed the last thing, it closely followed the series of reprints he had been busy turning out, and with which he would help to support himself. And there is a certain irony in the fact that "The Printer's Farewell" was subsequently to be one of the ten Strawberry Hill detached pieces that were themselves forged in the early nineteenth century: forgeries in which Kirgate had no part, but which unwittingly followed a practice he had begun.[125]

The Printing House Displayed

THE ARTIST Edward Edwards came to Strawberry Hill in 1781 to draw views that were to be engraved for the 1784 edition of the *Description*. His first commission appears to have been the completion of the watercolor by Thomas Sandby (1721–1798) of the Gallery, of which he prepared a smaller version for the engraver.[126] His association with the house lasted until 1784, when he and Walpole fell out over "what Mr. Walpole considered to be an overcharge for a cabinet made by a person recommended by Edwards"; this was the Beauclerk Cabinet, designed by Edwards to display drawings by Lady Diana Beauclerk.[127] One of Edwards's drawings was a watercolor of the Printing House, situated in the farmyard, with Kirgate walking in front of it with a portfolio under his arm; this is the primary image of the Printing House as built by Walpole in 1759, the most copied and the most often reproduced (see Figs. 24–27, discussed below). It was not engraved for the *Description,* not least because the *Description* contains no account of the Printing House in its survey of the grounds and garden buildings of Walpole's estate. Although the primary, it is not technically the first image of the Printing House, as there survives a very rough outline pen-and-ink

23 (Cat. 1.2)
Horace Walpole, *"View of Farmyard & Printing house at Strawberry hill,"* probably early 1780s. Pen and ink, 16.1 x 19.9 cm

sketch by Walpole, inscribed on the back in Kirgate's hand: "Sketch by Mr. Walpole (when his Hands were wrapt in Bootikins in a fit of the Gout) for a Print, before *that* by Edwards was made" (Fig. 23; Catalogue 1.2). It shows, crudely drawn and out of proportion, the farmyard barn in the center, with to its right the apparently conical, thatched roof of the ice house, and to the right of that a simplified, undersized version of the Printing House. Barn, Ice House, and Printing House are identified in pencil beneath, in Kirgate's hand; and at the foot of the sheet, Walpole has haltingly written in ink "View of Farmyard & Printing house at Strawberry hill." Given the close control exercised by Walpole over images of the house and grounds and collection, it is quite possible that this sketch may have been intended to prescribe in general terms what was required of Edwards.

The Edwards view appears in a variety of forms. At the Lewis Walpole Library there are three different watercolors, as well as various copies of the related aquatint by Edwards and Francis Jukes (1745–1812). The finished watercolor by Edwards has a neatly written title in capitals—"Printing House at Strawberry Hill"—in a wash-line mount and is dated 1784 (Fig. 24; Catalogue 1.3). Edwards prepared an outline etched plate of the view, which he initialed and dated "E. E. 1783"; and there is a second watercolor by him, on a sheet pulled from the proof etching before the aquatint process was employed (Fig. 25; Catalogue 1.4); the cloud formation is different, but it is otherwise very close to Fig. 24. The aquatint plate of the image, "Drawn & Etch'd by E Edwards. A. / Engrav'd in Aquatinta by F Jukes," was published on 1 February 1784 (Fig. 26; Catalogue 1.5). Finally, the third watercolor (Fig. 27; Catalogue 1.6), undated and by another hand, is a looser version than the Edwards watercolors, less carefully drawn and colored, with the holograph title beneath: "Farm Yard and Printing House belonging to Strawberry Hill." It may well have been copied from the aquatint. The aquatint was published as one of a pair by Edwards and Jukes, the other view showing the south front of the house (Fig. 28). Both aquatints were issued printed in brown ink, though some were printed in black.

The perspective chosen by Edwards is from the northeast and shows the barn to the left, the ice house to its right largely hidden by trees, and then the Printing House beyond. It is a simple brick building under a tiled roof, with an entrance door at the center of the north elevation; two small sash windows, one above the other, to the right of the door; and one large sash window to the left, with another large sash on the flank (eastern) wall. The cornice is represented by a simplified white band. In front of the building is the wooden paling bounding its front garden, and beyond that the brick wall that separated the side of the house from the Teddington road is lightly sketched. The fenestration defines the internal layout, which consisted of the composing room to the west, or right, of the front door, with Kirgate's chamber above, while to the left, or east, of the front door was the press room, a single-storied,

24 (Cat. 1.3)
Edward Edwards, *"Printing House at Strawberry Hill 1784,"* 1784. Pen and ink and watercolor, 17.9 x 23.8 cm (image), in a wash-line mount

25 (Cat. 1.4)
Edward Edwards, *"Printing House at Strawberry Hill,"* 1783. Etching and watercolor, 18 x 23.3 cm (image), in a wash-line mount

26 (Cat. 1.5)
Edward Edwards and Francis Jukes, after Edwards, "*Farm Yard and Printing House at Strawberry Hill,*" 1784. Etching and aquatint, 18.2 x 23.7 cm (image). Laid into Bull's extra-illustrated copy of the *Description* of Strawberry Hill

27 (Cat. 1.6)
Artist unknown, after Edward Edwards, "*Farm Yard and Printing House belonging to Strawberry Hill,*" after 1783–84. Pen and ink and watercolor, 18.3 x 23.9 cm (image)

28
Edward Edwards and Francis Jukes, after Edwards, *"South Front of Strawberry Hill the Seat of the Hon. H. Walpole,"* 1784. Etching and aquatint, 18.4 x 23.5 cm (image). Laid into Kirgate's extra-illustrated copy of the *Description* of Strawberry Hill

double-height space lit by the two large sash windows. It seems clear that the chamber was where Kirgate slept, as the third stanza of his "Printer's Farewell to Strawberry-Hill" opens with the couplet "Farewell! my PRINTING-HOUSE, farewell! / Where I no more shall calmly dwell." Walpole's letter to Mary Berry of 6 November 1795 provides further evidence of this arrangement:

> We had a tremendous tempest of wind this morning before five o'clock; it did not wake me till the close, though it has done me mischief. It has levelled the two tall elms in the meadow beyond the clump of walnut trees, and snapped two others short in the grove near the terrace; it ripped off the tiles from the corner of the printing-house, and Kirgate rose in a panic.[128]

The appearance of Kirgate's chamber is not known, but we do have a pair of views taken from it: watercolors by Silvester Harding (Fig. 29; Catalogue 1.18) and George Perfect Harding (see Fig. 80; Catalogue 1.19), the one copied from the other. The Silvester Harding version has inscribed beneath it, in Kirgate's hand, "View from the Chamber Window of the Printing House at Strawberry Hill." They show the

29 (Cat. 1.18)
Silvester Harding,
"View from the Chamber Window of the Printing House at Strawberry Hill,"
probably 1790s. Pen and ink and watercolor, 12.9 x 9.5 cm (image)

prospect to the west, looking from a window in the side wall of the Printing House, beyond the Teddington road, over farmland toward houses on the horizon to the left and a barn on the right.

There are two later versions of the Edwards image of the Printing House in the form of vignettes. The first of these (Fig. 30; Catalogue 1.7) was engraved by Philip Audinet (1766–1837) for the third volume of *The Bibliographical Decameron; or, Ten Days Pleasant Discourse upon Illuminated Manuscripts, and Subjects Connected with Early Engraving, Typography, and Bibliography* (1817) of the Reverend Thomas Frognall Dibdin (1776–1847). Consistent with Dibdin's bibliographical interests, the barn and ice house have been excluded as irrelevant, and the vignette is effectively a close-up view of the Printing House with Kirgate's figure prominent in the foreground. It is described in Dibdin's fulsome prose, addressing his reader in a typically discursive footnote, "thou hast it here, from an ancient etching, with the said Thomas Kirgate advancing from the office, with a proof sheet, towards the owner of the mansion."[129] The second, "The Printing House of Horace Walpole at Strawberry Hill. Engraved by William Collard from an Original Picture" (Fig. 31; Catalogue 1.8), is a slightly later

30 (Cat. 1.7)
Philip Audinet, after Edward Edwards, *Printing House at Strawberry Hill,* 1817. Etching and engraving, 9.8 x 10.5 cm (image). From Thomas Frognall Dibdin, *The Bibliographical Decameron* (London, 1817), 3: 450

31 (Cat. 1.8)
William Collard, after Philip Audinet, *"The Printing House of Horace Walpole at Strawberry Hill,"* ca. 1820s. Steel engraving, 9.8 x 14.9 cm (plate)

steel engraving copied from Audinet, but different in minor details, such as the posts in the front fence not projecting above the cross-rail, and the suggestion of foliage at the lower left. The Edwards view was reused as late as 1930 (see Fig. 79; Catalogue 1.9), being employed in the form of a zinc line cut illustrating *Horace Walpole and His Printing Office at Strawberry Hill,* a pamphlet issued by the Walpole Printing Office of New Rochelle (a press founded in 1929 that specialized in limited editions) as a sample of the quality of its printing.

The Audinet print can be usefully compared with two further vignettes engraved by him for Dibdin's *Bibliographical Decameron,* showing the printing houses of two other private presses, at Hafod and Auchinleck. The Hafod Press (Fig. 32) of Thomas Johnnes (1748–1816) was active from 1802 to 1809, and the Auchinleck Press (Fig. 33) flourished from 1815 to 1822, when the proprietor—Alexander Boswell, son of the biographer—was killed in a duel. Both these presses were of the generation after Strawberry Hill, Hafod printing Johnes's translations of Froissart, Monstrelet, Joinville, and De Brocquière in quarto, while at Auchinleck, printings were usually limited to forty copies, with an emphasis on early poetry. Neither could begin to match the

32
Philip Audinet, after Thomas Frognall Dibdin, *Printing House of the Hafod Press,* 1817. Etching and engraving, 7.8 x 11.8 cm. From Dibdin, *The Bibliographical Decameron* (London, 1817), 3: 362

33
Philip Audinet, *Printing House of the Auchinleck Press,* 1817. Etching and engraving, 6.3 x 9.9 cm. From Thomas Frognall Dibdin, *The Bibliographical Decameron* (London, 1817), 3: 454

longevity or variety of Strawberry Hill, but it is interesting that, as at Strawberry Hill, the printing house was situated in a cottage on the grounds. At Hafod it was in the hills a mile and a half away from the mansion, though at the time of Dibdin's visit it was occupied by a poor Welsh-speaking woman and her family, all traces of the press having disappeared. At Auchinleck it was, as Alexander Boswell explained, originally constructed for another purpose, and was "very pleasantly situated on the bank of a rivulet; and, although concealed from view by the surrounding wood, not a quarter of a mile from my house."[130] In both cases the printers employed were fortunate compared to the printer of the irascible, debt-ridden Sir Thomas Phillipps (1792–1872) at the Middle Hill Press, initially housed above the scarp of the Cotswold hills in the damp and windswept fastness of Broadway Tower (Fig. 34), where, according to Phillipps's agent, "the Windows are so bad Broken that the wet when it rains floods every Room, & runs through the floors & Ceilings so that the Plaistering is coming down in many places."[131] The press was subsequently moved to the basement of Middle Hill itself, but conditions there were far from ideal. At another private press of the next generation, that of Sir Egerton Brydges (1762–1837) at Lee Priory, in Kent (active 1813

34
Artist unknown, *Broadway Tower,* 1834. Woodcut, 5.4 x 7.5 cm. From John Martin, *A Bibliographical Catalogue of Books Privately Printed; Including Those of the Bannatyne, Maitland and Roxburghe Clubs, and of the Private Presses at Darlington, Auchinleck, Lee Priory, Newcastle, Middle Hill, and Strawberry Hill* (London, 1834), 441

35
John Preston Neale, *"Lee Priory, Kent,"* ca. 1825. Pen and ink and watercolor, 8.4 x 12.4 cm. Private collection. The watercolor was subsequently engraved for Neale's *Views of the Seats of Noblemen and Gentlemen in England, Wales, Scotland, and Ireland,* 2nd ser., vol. 2 (1825).

36 (Cat. 3.1)
Edward Edwards, finished by John Hall, *"T.K."* (Thomas Kirgate), 1784–90. Etching, 12.5 x 8.3 cm (plate). Bound into Kirgate's second copy of the *Description* of Strawberry Hill

to 1823), the literary and antiquarian output had at least some resemblance to that of Strawberry Hill, as had the house itself, extended and adorned by James Wyatt for Walpole's correspondent Thomas Barret (ca. 1743–1803) in the late 1780s, and celebrated by Walpole as a "child of Strawberry prettier than the parent" (Fig. 35).[132] Unlike Strawberry Hill, the press was manned by a compositor and a pressman; Brydges provided the copy, largely old and out-of-the-way English texts, and the printers produced small printruns and kept such profits as they made. The press was housed in a vacant room, at the extremity of the offices.[133]

Edwards not only painted what appears to be the earliest effective image of the Printing House; he also painted the earliest portrait of Kirgate. The original, which was exhibited at the Royal Academy in 1784 as exhibit 152, is now lost, though we know that Walpole saw it, as he annotated his copy of the catalogue by inserting after the printed word "Portrait" the words "of T. Kirgate, Mr. Hor. Walpole's Printer."[134] What survives, however, is the etching of it, a head and shoulders of Kirgate looking to the right, the finished version titled "T.K." and inscribed as painted and etched by Edwards, and finished by J. Hall (Fig. 36; Catalogue 3.1). It is an engaging face, with

37 (Cat. 3.2)
G. Barrett, after Edward Edwards, *"Thos. Kirgate Lord Orford's Printer at Strawberry Hill,"* after 1791. Etching, 12 x 7.6 cm

38 (Cat. 3.3)
William Collard, after G. Barrett, *"Thomas Kirgate Printer to Horace Walpole at Strawberry Hill,"* ca. 1820s. Steel engraving, 24 x 14.8 cm (plate)

a sympathetic, inquiring expression beneath a receding hairline. There also exist two later engravings after it, both undated: one, slightly coarser, published by G. Barrett (Fig. 37; Catalogue 3.2); and a weaker, steel-engraved version (Fig. 38; Catalogue 3.3) by William Collard (1792–1847)—who also re-engraved the Audinet vignette of the Printing House with similarly styled lettering, presumably at much the same time (Fig. 31).[135]

The story of Edwards's plate is explained by the artist himself, in a letter addressed to Richard Bull, laid into Bull's extra-illustrated copy of the *Description:*

> Mr. Edwards Compliments to Mr Bull & knowing that he collects every print &c. that relates to Strawberry hill begs his acceptance of the inclosed portrait of which E can say nothing but that it is & ever will be scarce as Mr Walpole thought fit to be offended at its being done & therefore its publication will be suppressed Mr Walpole has no more than two proofs & E has only printed off a few impressions on purpose to oblige some of his curious Friends & acquaintance. E must particularly beg that it *may not be lent or parted* with so that it may be copied.
>
> No. 36 Windmill Street Rathbone Place
> Decem 18 1784[136]

Immediately beneath the letter is the print itself, in unlettered proof state, inscribed in manuscript "Mr. Walpole's Printer"—the same proof state that Walpole laid into his manuscript *Journal of the Printing-Office,* which he labeled by hand "E. Edwards p. & sc. 1784." The print was subsequently finished by J. Hall by adding the lettering and was published by Edwards in a collection of etchings in 1790, once he had not only

39
Edward Edwards,
"Strawberry-Hill" (entrance to Strawberry Hill), 1781.
Pen and ink and watercolor on laid paper with wash-line mount, 19 x 13.5 cm (image). Laid into Walpole's extra-illustrated copy of the *Description* of Strawberry Hill

ceased to work for Walpole, but also felt free of the restrictions that Walpole placed on what images of Strawberry Hill were circulated.[137] There is no evidence as to exactly when in 1784 they fell out over the cost of the Beauclerk Cabinet, but it would not be surprising if it were before that year's Royal Academy exhibition, where the painted portrait was displayed. For it is a curious fact that despite the extraordinary wealth of drawings, paintings, and prints of Strawberry Hill, this is the only image we have of a member of Walpole's household circulated during his lifetime—and this Walpole wanted suppressed. His need to record the house, the grounds, the garden buildings, the treasures of his collection, did not extend to providing any record of the servants who worked there. Where these views do show people—the servant at the front door in Edwards's watercolor of 1781 (Fig. 39); Kirgate walking before the Printing House, again by Edwards (Figs. 24–27); and the servant with stick and portfolio emerging from the Cottage in the Garden (quite possibly Kirgate, as part of Walpole's library was kept in the Cottage) in the watercolor by J.C. Barrow (Fig. 40)—they are not in any sense portraits, but are merely incidental to the pictorial record: Walpole's staff in the role of staffage. Walpole seems to have preserved no portrait, for example, of his Swiss valet, Philip Colomb, who served him from about 1767; not for Walpole the servant's hall of Erddig, lined with generations of the family's faithful servants.[138] Our pictorial record of Strawberry Hill, rich, varied, and textured as it may be, is not emotionally enriched by any sense of a community living behind that "embattled wall."[139]

It is consistent with this interpretation that the most detailed portrait we have of Kirgate, drawn by Silvester Harding in 1794 and engraved by Andrew Birrell, was not

40
Joseph Charles Barrow, "*The Cottage*," 1791. Watercolor, 21.5 x 28.4 cm. Laid into Walpole's extra-illustrated copy of the *Description* of Strawberry Hill

engraved and published until 1798, the year after Walpole's death. Titled "Thomas Kirgate, Printer at Strawberry=Hill. With a View of the Printing House," it shows Kirgate in closely observed profile, with the facade of the Printing House beyond (Fig. 41; Catalogue 3.4). It clearly displays the disposition of doors and windows, the layered cornice, the pantiled roof, and in front of the building a neat, low fence that separated Kirgate's garden from the farmyard. It also illustrates the position of the chimney, rising above the roof to the right of the front door, providing fireplaces to heat the composing room and Kirgate's chamber above. Kirgate's face is more deeply lined than in the Edwards etching, reflecting the passage of twelve or so years that separate the two images. The figure in profile is detached from the background view, being very much a studio portrait of the sitter, to which the artist added as a backcloth the view of the Printing House, with a prominent tree behind Kirgate to complete the composition. There survives a trial proof state of the print, before the engraved title and some of the detailing were added (Fig. 42); this is inscribed in pencil "but 2 impressions taken Mr. Baker [the collector George Baker (1781–1858)]." There is also another version of the image that appeared nearly ninety years later, when extra-illustration was at its most florid and self-indulgent: in a copy of L.B. Seeley's *Horace Walpole and His World* at the Lewis Walpole Library, expanded in 1884–90 into twelve folio volumes and dressed in decoratively gilt red morocco, is a painstakingly exact pen-and-ink copy of the Birrell print (see Fig. 83; Catalogue 3.5), initialed "D.E.C." for D.E. Cronin, who appears to be David Edward Cronin (1839–1925), an American lawyer and illustrator active as an illustrator of books in the 1880s and 1890s, who also supplied the set with verdant Victorian watercolors.[140] It is identical, save that it does not quite capture the animation that Harding gave to Kirgate's expression. This pair of images does record, however, that sense of identity and sympathy between printer and Printing House that was in 1797 put asunder.

But this is to move ahead. Chronologically, the next image of the Printing House after Edwards's watercolor of 1784 is a recently discovered, anonymous amateur watercolor, a view from the common field north of the estate, taking in Walpole's Round Tower at the center, and to the right, shielded from the Teddington road by a brick wall, Walpole's farmyard with, at the far right, the Printing House (Fig. 43; Catalogue 1.10). The picture can be dated between 1776, when the Beauclerk Tower visible to the left of the Round Tower was erected, and 1790, when the Offices were built behind the farmyard buildings. Its interest for our purposes is twofold: it is the only image that shows the exterior of the western flank wall of the Printing House, with two ground-floor windows lighting the composing room and two windows above lighting Kirgate's chamber; and it is also the only image giving a general, if distant, view of Walpole's farmyard before the barn was demolished to make way for the access to the Offices. It can be read in conjunction with a map by Claude Joseph Sauthier, drawn in 1787 to show the Duke of Northumberland's holdings in the manor of Isleworth (Fig. 44; Catalogue 2.1). Sauthier's map predates the Carter maps (Figs. 11 and 12); and like the anonymous watercolor, it shows the farmyard before the Offices were erected. As in the watercolor, the Sauthier map shows four separate buildings southwest of the house, of which the Printing House is the furthest west, adjoining the Teddington road, though its shape is given as a less elongated rectangle than in the other images and plans. The building immediately to the left of the Printing House in the watercolor would appear to be a version of the barn shown at the left of the Edwards watercolor (Fig. 24); but although it is in the same position and of similar proportions, with a similar pitched tiled roof, its orientation has been changed: the gable end faces west in the anonymous watercolor and north in the Edwards view. Of the other two buildings shown in the farmyard, it is possible that the most easterly, on the far left of the farmyard, is the greenhouse ("E" in Walpole's copy of Carter's map of the estate).

We now turn from a quietly anonymous amateur artist to a positively vociferous professional. John Carter was a draftsman, architect, and passionate defender

S. Harding ad vivum delin. 1794. A. Birrell sculp.

THOMAS KIRGATE,

Printer at Strawberry-Hill.

With a View of the Printing House.

Published in 1798.

41 (Cat. 3.4) *opposite*
Andrew Birrell, after Silvester Harding, "*Thomas Kirgate, Printer at Strawberry=Hill. With a View of the Printing House,*" 1798. Etching, engraving, and stipple, 12.4 x 11 cm (image). Laid into Bull's extra-illustrated copy of the *Description* of Strawberry Hill

42
Andrew Birrell, after Silvester Harding, *Thomas Kirgate, Printer at Strawberry=Hill. With a View of the Printing House,* 1798. Etching and engraving, proof before letters and stipple, 12.4 x 11 cm (image)

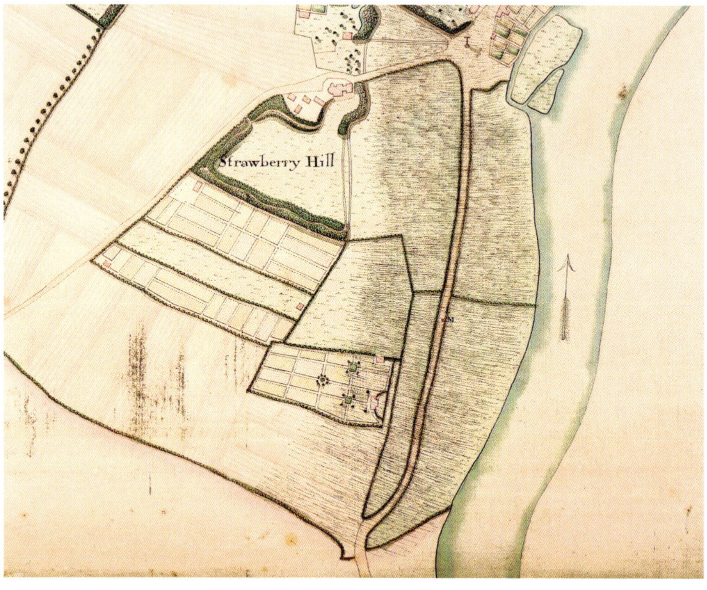

43 (Cat. 1.10)
Artist unknown, *Strawberry Hill, Farmyard, and Printing House from the Common Field,* 1776–90. Pen and ink and watercolor, 21.8 x 30.9 cm. The Friends of Strawberry Hill

44 (Cat. 2.1)
Claude Joseph Sauthier, Detail of *"A Map of the Manor of Isleworth in the County of Middlesex belonging to his Grace the Duke of Northumberland,"* 1786–87. Pen and ink and watercolor on paper backed onto linen, 280 x 231 cm. Collection of the Duke of Northumberland

of the Gothic from the depravations of modern "improvement," in particular the improvements dispensed by the architect James Wyatt (1746–1813), against whom he conducted a form of guerrilla warfare in the pages of *The Gentleman's Magazine* from 1797 until his death. Carter saw Wyatt as the heir to Batty Langley (1696–1751), the publisher of decorative but archaeologically incorrect Gothic confections, and castigated him repeatedly as "The Destroyer" for his innovation and lack of authority in the restoration of Gothic cathedrals.[141] Yet from 1780 for about ten years, Carter had been employed by Walpole as one of the artists recording Strawberry Hill, where for all the use of specific examples of Gothic detailing to create a fireplace or a ceiling, the spirit of innovation was exercised in a manner not wholly dissimilar to that which Carter was to lambast in his later journalism. His watercolors of Walpole's interiors, such as the Holbein Chamber, the Tribune, the Armoury, and the Library, have been much reproduced as capturing that combination of intricacy and lightness that, for all the areas of Gothic gloom, characterized much of the house; and he seems to have embraced the Walpolian fantasy with his detailed, bravura watercolors illustrating scenes from Walpole's *The Castle of Otranto* (1765).[142] Working both for Walpole and for Bull, Carter also drew numerous and valuable images of items from Walpole's collection. At the Huntington Library is probably the most important of the extra-illustrated copies of the *Description* outside the Lewis Walpole Library, the extensively extra-illustrated copy compiled by Carter, a folio with the text inlaid and adorned on page after page, recto and verso, with his drawings and sketches and finished watercolors, and with a title page printed at the Press, "Drawings and Sketches, of the Gothic Mansion at Strawberry-Hill, Twickenham, the Seat of the Honourable Horace Walpole; and of several of the Antiques and Curiosities in his Collection: taken, by permission, by John Carter, Draughtsman and Designer of Gothic Architecture, 1788. Printed at Strawberry-Hill, by T. Kirgate, 1788."[143] In this volume lie five sketches of the Printing House (Figs. 45 and 48–49).

The first of these is a loosely drawn sketch of the building from the northeast (Fig. 45; Catalogue 1.11), similar in perspective to the Edwards view. It provides little further evidence, other than the suggestion on the left of the entrance to Walpole's original ice house. But the four other Carter images in the volume are much more suggestive, for they are the preparatory sketches to his view of the interior of the Printing House. There are two versions of the finished watercolor of this interior, one at the Lewis Walpole Library (Fig. 46; Catalogue 1.14), laid into Bull's copy of the *Description,* and the other at the Huntington Library (Fig. 47; Catalogue 1.15), laid into Bull's volume of Strawberry Hill detached pieces. In addition, there is in Kirgate's elephant folio Strawberry Hill scrapbook a separate watercolor of the interior, probably by George Perfect Harding (see Fig. 54). Together these are the only general views we have of the interior of an English eighteenth-century or early-nineteenth-century private press.

45 (Cat. 1.11)
John Carter, *"Northeast view of the Printing-house, at Strawberry-hill,"* probably mid-1780s. Pencil, 21.1 x 18.8 cm (irreg). Laid into Carter's extra-illustrated copy of the *Description* of Strawberry Hill. The Huntington Library, Art Collections, and Botanical Gardens

As such they deserve study for what they tell us about how such a press was arranged and equipped.

The Lewis Walpole Library version of the Carter watercolor (Fig. 46) has been reproduced a number of times—most prominently as the frontispiece to Hazen's *Bibliography*—but even so it has never been analyzed. The watercolor shows the composing room of the Printing House with the slight figure of Kirgate (who was described as "small in his person") setting type.[144] On the right, on the near side of the fireplace, is a doorway opening onto the press room, where Kirgate's assistant is working the press. The view is taken from the rear of the composing room: the window at the far end, over Kirgate's shoulder, is the low sash window to the right of the front door in the Edwards view (Fig. 24), while the two low sashes to the left are those in the western flank wall of the building, as seen in the watercolor taken from across the common field (Fig. 43). The perspective is distorted, making the room look deeper than it really was—the doorway on the right was in fact at the rear of the internal dividing wall. The stairs would seem to have been in the press room, immediately inside the front door, set against the internal wall and so not visible in Carter's view. Kirgate's composing frame is placed between two windows of the composing room for maximum light, with—as recommended in Moxon's *Mechanick Exercises*—the frame placed so that the light comes in on the compositor's left hand, "for else his Right hand plying between the Window-light and his Eye might shadow the *Letter* he would pick up." To the extreme right, in the press room, one can see that the press is lit by the tall sash window in the eastern side wall of the building; it would also have been lit by the tall sash in the north facade—shown in Edwards's exterior view—again complying with Moxon's advice that the press should be well lit with a "clear, free and pretty lofty Light," and placed so that "the Light may fall from a Window right before the *Form* and *Tinpan:* And if scituation will allow it, on the North-side the Room, that the *Press-men,* when at their hard labour in *Summer* time, may be the less uncommoded with the heat of the *Sun*."[145] The other advantage of a north-facing press room was to prevent the sun drying the paper as it was being printed.

The basic outline of the composing room is set out in a simple pencil sketch laid upside down in Carter's extra-illustrated copy of the *Description* (Fig. 48; Catalogue 1.12), which provides the framework for both versions of the watercolor. The other three pencil sketches in Carter's *Description* volume are figure studies on a single sheet (Fig. 49; Catalogue 1.13). One shows Kirgate at his composing frame, and one is of his assistant working the press, both in the poses used in the Lewis Walpole watercolor (suggesting that it, rather than the Huntington version, is the primary image). Beneath them is a third pencil sketch, not used in the watercolor, showing Kirgate sitting with two young men (presumably his assistant and another) around a small circular table, taking tea: "No more in Conversation free, / Enjoy my Friend and sip my Tea," as he was to lament in "The Printer's Farewell to Strawberry-Hill." The chair in which Kirgate is sitting, with its distinctive curved shape, is clearly visible in the watercolor, situated behind the second composing frame to which Kirgate's back is turned. The circular table is immediately in front of it.

The fact that the watercolor shows an assistant working with Kirgate helps us to date it. Walpole usually employed just the one printer, and the slow pace of printing without an assistant had its frustrations. As early as July 1757, when William Robinson was working on Gray's *Odes,* the impatient Walpole had complained that "The Press goes on as fast as if I printed myself"; and in 1773 he was forced to acknowledge to William Mason, in accepting that the Press was not best suited to producing Mason's life of Gray, "that the pace of my single printer, who has not even an aide-de-camp or devil, is so wondrous slow that your work would not be finished in this century."[146] But on two occasions, Walpole did employ an assistant. The letter of engagement of August 1759 from Thomas Farmer, his fourth printer, contained an agreement by the printer to "instruct in the art of printing, during the said two years, any young man

46 (Cat. 1.14)
John Carter, *"Inside of the Printing-House at Strawberry-hill,"* probably 1785. Pen and ink and watercolor, 14 x 21.9 cm (image). Laid into Bull's extra-illustrated copy of the *Description* of Strawberry Hill

47 (Cat. 1.15) *opposite*
John Carter, *"Interior of the Printing house at Strawberry hill, with a Portrait of Mr. Kirgate the worthy Printer employ'd by Lord Orford at the time of his death, and for many years previous thereto,"* 1785 or later, the inscription provided after 1797. Pen and ink and watercolor, 15.6 x 22.1 cm (image). Laid into Bull's *Strawberry Hill Detached Pieces.* The Huntington Library, Art Collections, and Botanical Gardens

Interior of the Printing house at Strawberry hill, with a Portrait of Mr. Kirgate the worthy Printer employ'd by Lord Orford at the time of his death, and for many years previous thereto.

you shall appoint to attend upon me."[147] Joseph Forrester had already been taken on as apprentice in the previous month, this being the time that the Press was engaged in the substantial undertaking of printing Lucan's *Pharsalia,* "a work almost beyond our compass."[148] And in July and August 1785, Edward Yardley was taken on to help Kirgate (who spoke no French) print the translation of the *Essay on Modern Gardening* by the Duc de Nivernois, and it appears that Yardley is the assistant shown and 1785 the date of Carter's watercolor.[149]

Returning to Kirgate standing before his composing frame: he has his page of printer's copy lying on the upper case, where the uppercase or capital letters were kept, the lowercase letters being kept in the lower case below. To his left, the shelf extending from the side of the frame was probably used for imposition, the transfer of the lines of type from his composing stick to the chase, which may be the item visible on the shelf. The sheet of copy can be seen clearly in a watercolor portrait of Kirgate standing with his back to the frame, now in the British Museum (Fig. 50; Catalogue 3.6). Kirgate is holding out his composing stick in his left hand. As in the Carter view, Kirgate is not using a visorium, a small stand to hold his page of copy, but has simply laid the copy on the upper case. The size of the boxes of type in the lower case varies, those in the center being larger; these were reserved for the most frequently used letters. The watercolor has beneath it Kirgate's signature, with his distinctive "K," but is not signed by the artist. However, there is a rare print taken from the watercolor, an etching with stipple engraving inscribed "S. Harding delt. E. Harding Junr. Sculp." (Fig. 51; Catalogue 3.7), so we know that Kirgate's friend Silvester Harding painted

48 (Cat. 1.12) *opposite*
John Carter, *Interior of the Printing House,* probably 1785. Pencil, 13.5 x 23.4 cm (irreg.). Laid into Carter's extra-illustrated copy of the *Description* of Strawberry Hill. The Huntington Library, Art Collections, and Botanical Gardens

49 (Cat. 1.13)
John Carter, *Three Figure Sketches inside the Printing House,* probably 1785. Pencil, 20.1 x 15.2 cm (irreg.). Laid into Carter's extra-illustrated copy of the *Description* of Strawberry Hill. The Huntington Library, Art Collections, and Botanical Gardens

Thos Kirgate

50 (Cat. 3.6) *opposite*
Silvester Harding, *Thomas Kirgate,* 1790s. Pencil and watercolor, 16 x 13.2 cm (image). The Trustees of the British Museum

51 (Cat. 3.7)
Edward Harding, after Silvester Harding, "*Thos. Kirgate. Printer at Strawberry Hill,*" no later than 1796. Etching and stipple engraving, 16.8 x 13.9 cm (image)

it and that his elder son, Edward, engraved it. The print's scarcity is explained by the note inscribed by Kirgate to the collector George Baker on the copy at the Lewis Walpole Library: "The Plate destroyed, being unlike the Person. For George Baker, Esqr. with Kirgate's respects."[150]

This print of Kirgate is not without its mysteries. If the composing stick in the watercolor is looked at upside down, the word "Anecdotes" in capitals can be made out; this is much clearer in the print. Furthermore, the page of copy, blank in the watercolor, is in the print made up as a title page for the *Anecdotes of Painting,* reading "ANECDOTES / of / PAINTING ENGLAND / [triple rule] / Mr GEORGE VERTUE / [rule] / By Mr HORACE WALPOLE / [double rule] / VOL I / STRAWBERRY HILL / PRINTED BY / THOMAS KIRGATE." The irony is that Kirgate printed only the fourth and final volume of the *Anecdotes,* Thomas Farmer being named as printer on the title page of the first two volumes, and William Pratt having completed the first two volumes after Farmer ran away, and printed the third. Additional confusion is provided by an early-twentieth-century photogravure of untraced origin (see Fig. 84; Catalogue

3.8) showing the same portrait. The image is the same, but the lettering is missing from the composing stick and the page of copy, suggesting that the photogravure may have been taken from a now lost proof of the print, before the lettering was added.

And what of the type lying in the trays on the frame? Toynbee was able to prove that in 1768 and 1769, Walpole bought English and Pica type from the letter founder William Caslon the younger (1720–1778).[151] Hazen adds that the English type was bought for the abortive 1770 printing of Walpole's *Works* in quarto, and that on the surviving copies "the relatively clean, unbroken type increases the effectiveness of the pages."[152]

Turning again to the Carter watercolor, it has other things to tell us about the facilities and operation of the Press. Between the two windows on the side wall to the left is a table, at the foot of which is a large rectangular package, possibly of printed books waiting to be sent out, but more likely of paper for the press (there appears to be another inside the door of the press room, on the floor to the left). The only contemporary record we have of Walpole's purchase of paper is from the very first year in which the Press operated, a letter from Walpole to his deputy Grosvenor Bedford (d. 1771, the father of Charles Bedford) of 9 November 1757:

> I must desire you will speak to Mr Tonson to send me another parcel of paper for my printing, but I wish he would order it to be carefully examined, because in the last parcel there were several thin sheets interspersed. As I shall be in town the end of this week, I shall be glad if he would have the bills made up of the expenses of the press, etc., that I may pay them.[153]

Tonson was Jacob Tonson, who in July of that year, as we have seen, had been summoned by Walpole with Robert Dodsley to witness the opening of the press. Hazen made a careful study of the watermarks on which the books and detached pieces were printed, and used it to distinguish between the original productions of the Press and Kirgate's reprints, but also noted how little information we have on Walpole's sources of paper.[154]

Other equipment in the composing room includes, on the chest of drawers at the far left, an inking roller or mallet for knocking down the type once it has been locked up in the chase, and, leaning against the chest, what appears to be a chase with forme consisting of two pages locked up in it. On the table between the windows is a variety of objects including a jug and what may be a quill pen, together with a pile of paper being pressed flat after printing. Turning to the press room visible through the open doorway, we see the wooden common press being worked by Kirgate's assistant. He is working at half press, which is to say that he is carrying out both operations of beating the forme (inking the type with leather-covered ink balls) and pulling the bar (which pressed the paper down onto the wet ink). It was more common for two workmen to share the tasks of beating and pulling, which was known as working at full press. For the rest of Kirgate's working life he would have worked alone, at half press; six years before first employing Kirgate, Walpole had made clear that he expected his printer to "both compose and take off" — that is, compose or set the type, beat, and pull.[155]

The press at which the assistant is working is partly obscured by the assistant himself and by the table behind him, but it is possible to see the two upright side pieces, or cheeks, of the press, joined across the top by the horizontal section of wood called the cap. The thick length of wood rising diagonally upward from the nearest cheek of the press was a stay, securing the press to the wall; because of the pull on it by the pressman, Moxon advised that the press be "fastned with Braces overhead and on its Sides, as well and steddy as under foot."[156] Above the press hang the printed pages drying on racks, being thin deal board battens. To the assistant's left is the star-shaped handle of a second press; this was the rolling press, used to print engravings, which required heavier pressure than type. It was on this press that the prints illustrating the *Anecdotes of Painting* and the *Description* of Strawberry Hill, as well as the decorative vignettes that were used for some of the title pages, would have been printed; and it was this press that was out of order when in 1789 More's *Bishop Bonner's Ghost* was being printed.[157]

The Press initially employed two vignettes, described by Walpole as fleurons, the larger one with a shield designed by Bentley in 1757 (which was used on the title pages of Gray's *Odes* [Fig. 5], Hentzner's *Journey into England,* and Walpole's *Fugitive Pieces*), and the smaller one, designed by Müntz in 1758 (used on the title page of the *Royal and Noble Authors* [Fig. 6] and Whitworth's *Account of Russia*). They do not appear on the title pages of the books produced by the Press after 1758, presumably because they are early views of the east end of the house that were out of date by the time the Gallery and Round Tower were added in 1759–62.[158] They were both engraved by Charles Grignion (1717–1810).[159] Grignion provided others of the engraved decorations of the Press, in particular the frontispiece and coin of Queen Elizabeth for the *Royal and Noble Authors* and the two vignettes for Lucan's *Pharsalia;* the bravura title page decoration of weapons and torches against the background of Strawberry Hill; and the vignette of bust, books, and vine leaves under the letterpress of the address to the reader. In addition, Müntz drew and engraved the vignette portrait of Magliabechi for Spence's *Parallel.* The title page of the *Life* of Lord Herbert of Cherbury marks the first use of a framed, rectangular view of the house from the southeast, showing the Gallery and Round Tower; this was the print on which Kirgate experimented with printing in two colors (Fig. 7), and it was used again in the Countess Temple's *Poems,* the Honorable Richard Fitzpatrick's *Dorinda, A Town Eclogue,* and *Bishop Bonner's Ghost.*[160] Walpole pasted into a page of his extra-illustrated copy of the *Description* a selection of the vignettes used by the Press (Fig. 52). These include the larger Strawberry Hill vignette (center top), the smaller vignette (third down on the right), the later rectangular view (third down in the center) with immediately beneath it the Walpole crest from the title page of the first two volumes of the *Anecdotes of Painting,* the title page vignette (bottom right) and further vignette (bottom left) from the *Pharsalia,* and the portrait of Magliabechi from Spence's *Parallel* (third down on the left). The first and third versions of Walpole's bookplate appear next to bottom and bottom in the center column.

Carter's other interior view of the Printing House (Fig. 47), now at the Huntington Library, is at first glance almost identical to the Lewis Walpole Library version (Fig. 46), but on more detailed inspection a number of variations appear, showing that the one is not a mere copy of the other. The Huntington version has a neatly written title, "Interior of the Printing house at Strawberry hill, with a Portrait of Mr. Kirgate the worthy Printer employ'd by Lord Orford at the time of his death, and for many years previous thereto," added after Walpole's death in 1797. It is not clear when the picture itself was painted, but as noted above, the Huntington version is likely to be the later one. The most immediate difference between the two watercolors is the direction of the boarding of the floors, which in the Lewis Walpole version runs from back to front of the two rooms, toward the viewer (as structurally one would expect, so the supporting beams are at right angles to the fireplace wall), but in the Huntington version runs across the rooms. In the composing room, the Huntington version lacks the wrapped parcel of paper on the floor near the table, as well as the chase resting against the chest of drawers; and it has a slightly different arrangement of items on the table. The pictures on the wall differ, Kirgate looks up rather than down, and there is a piece of clothing or material thrown over the back of Kirgate's chair, a shovel to the right of the fireplace, and a different number of items hanging from the ceiling. The purpose of these is not clear, though they may have been glass fly-catchers, to keep flies away from paper that had been dampened for printing.[161] In the press room, there is a different disposition of items on the floor, and the Huntington picture lacks the stay securing the press, but it does have more clearly visible on the nearside cheek of the press what may be a ball rack, used for storing the inking balls.

One element absent from both images is any evidence of bookbinding; there has never been any suggestion that the books printed at the Press were bound on the premises. One would expect that the printed sheets, folded or not, would have been sent to the bookseller for binding, at least in the case of those titles printed in sufficient numbers to be distributed by London booksellers such as Bathoe. The pamphlets printed by the Press, however, if not rebound are found sewn in a rather amateur way into drab blue wrappers, stitched through the margin, with the thread

52
Strawberry Hill Press vignettes and two of Walpole's bookplates. Laid into Walpole's extra-illustrated copy of the *Description* of Strawberry Hill

53
Binding of front wrapper, *Miscellaneous Antiquities; or, a Collection of Curious Papers... Number 1.* Strawberry Hill Press, 1772

and its knot visible on the outside of the upper wrapper. It is quite possible that pamphlets in that format (particularly those printed in small numbers) were folded, stitched, and sewn on the premises, or at least locally. Even copies of the *Miscellaneous Antiquities* that survive in wrappers are stitched and sewn in this manner, despite a printrun of five hundred for sale (Fig. 53). The copies of the *Description* distributed after Walpole's death by his executors were also issued sewn into wrappers, inscribed on the wrapper "Mrs. Damer and Lord Frederick Campbell have pleasure in executing the directions of the late Earl of Orford."[162] Generally, though, buyers or recipients of the Press's books would have commissioned their own bindings, and hence the great variety of binding styles we see today in the books from the Press.

The third watercolor of the interior of the Printing House, attributed to George Perfect Harding, differs from the Carter images in a number of ways (Fig. 54; Catalogue 1.16). The viewpoint is almost identical, but slightly to the left; it is an independent image, not merely a copy of the Carter. It is clearly less accomplished and less sophisticated, but it is also clearly later in date. The composing room is empty, and there is no evidence of active printing; we are not able to see through the doorway to the press room itself. We can, however, now see that the frame to the right of the front window of the composing room has beneath it a cupboard, which would have contained the trays of type, and some of the floor-to-ceiling shelving in the corner behind it might have been used for storing paper. There is a locked-up chase for a quarto book leaning against the frame to the left, but this seems to be no more than a reference to past activity. The surfaces are almost bare, the floors are cleared, the hanging

54 (Cat. 1.16)
Attributed to George Perfect Harding, *Interior of the Printing House,* ca. 1800. Pen and ink and watercolor, 14.5 x 21 cm. Laid into Kirgate's elephant folio collection of Strawberry Hill images. The Lord Waldegrave of North Hill

items have been removed from the ceiling, and Kirgate's chair is drawn back against the wall to the right; the only domestic touch is the kettle to the side of the fireplace. The image has a slightly naive quality, with something of the naif's sense of otherness, detachment, and interior vacuum. But principally, it is a record of a disused space, a recollection of past activity; for it seems that by the time this watercolor was painted, Walpole was dead, and Kirgate departed.

Neglected and Forgot

Under walpole's will, a life interest in Strawberry Hill passed to the daughter of his cousin General Conway. The redoubtable Anne Seymour Conway had survived her husband, the spendthrift John Damer, who in 1776 committed suicide in a tavern; thereafter untempted by marriage, she devoted herself to sculpture, in which Walpole more than shared her own high opinion of her abilities (Fig. 55).[163] She was Walpole's residuary legatee and executrix, and was given £8,000, together with a further £2,000 to maintain the house.[164] She duly moved into Strawberry Hill with her mother, Lady Ailesbury; and with the house came its collections—which Walpole's will had directed should be treated as heirlooms—as well as the Printing House. The strength of Walpole's vision of Strawberry Hill as the home of a living collection is expressed in his efforts to secure the succession to the estate and its contents; it is also more particularly shown in one of the codicils to his will, in which he stipulates that the "prints, books of prints, and incomplete numbers of prints" in a trunk at his house in Berkeley Square should be added to the heirlooms at Strawberry Hill so that his heirs could, if they wished, "complete the books of prints in my said Round Library there as successive numbers of such works shall be published."[165]

Such aspirations for the further growth of the collection were not to be fulfilled. The heirlooms were preserved as a slumbering collection, still shown to visitors, but not expanded. The Printing House, meanwhile, entered into a period of change and decline. Anne Damer had no need of a printer, and Kirgate's services were promptly dispensed with. She had a taste for amateur theatricals, and when in 1800 she had printed a Prologue and Epilogue for a production she mounted there, they were not printed at Strawberry Hill, as there was no longer a press there.[166] It has already been noted that from about 1790 Kirgate had undertaken some printing on his own account; one example of this is his own address card, "Thomas Kirgate, Printer, Strawberry-Hill, Twickenham, Middlesex" (Fig. 56). He also printed at the Press a card for his friend, "Silvester Harding, Miniature Painter, and Engraver, No. 127, Pall Mall" (Fig. 57); he now printed one more for himself, recording his change of address to Harding's shop and his new vocation: "Thomas Kirgate, Late Printer at Strawberry-Hill, Print and Bookseller, No. 127, Pall-Mall" (Fig. 58).

55
Richard Cosway, *Portrait of Anne Seymour Damer,* 1790. Graphite with watercolor, 22.8 x 14 (visible). The Trustees of the Waldegrave Chattels Trust

Kirgate moved from Strawberry Hill first to Pall Mall, where Harding had his shop, and then to a house in Vine Street, Piccadilly, where he opened a book and curiosity shop. He was active in disseminating Strawberry Hill material, and his invoice of January 1798 to the Strawberry Hill collector and bibliographer George Baker shows him in the role of book and print dealer charging £6-0-6 for ten items, mostly Strawberry Hill-related (Fig. 59); Lewis in *The Forlorn Printer* reproduces another invoice from the following month in which Kirgate charged Baker £24 2s. 6d. for twenty Strawberry Hill Press and related items (some no doubt his own reprints).[167] His friend Silvester Harding's bookshop in Pall Mall, and before that in Fleet Street, was a center for circulating books from the Press and images, particularly historical portraits, from Walpole's collection. In this he was aided by his artist sons, George Perfect Harding and to a lesser extent Edward Harding, who died young; we have already seen that Silvester Harding supplied R.P. Cruden, and there were many other collectors eager to obtain Strawberry Hill material, whether the imprints of the Press, or prints and watercolors with which they could extra-illustrate their own Strawberry Hill titles. A glimpse of this world of print collecting, and of the extra-illustrator's desire to complete his volume, can be obtained from Kirgate's letter (year unknown, but the two other letters surviving with it are dated 1801) to Charles Bedford:

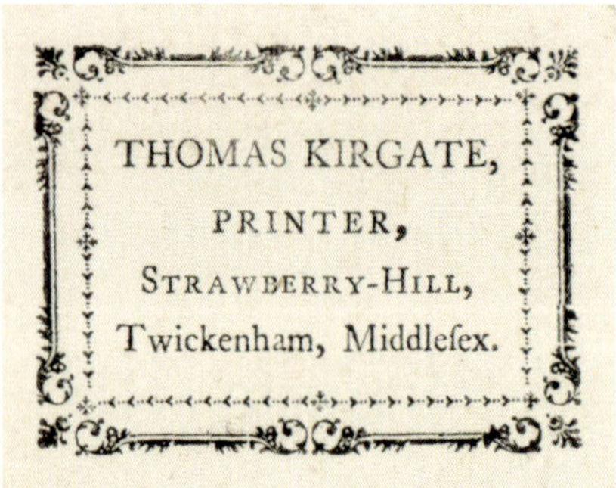

THOMAS KIRGATE,
PRINTER,
STRAWBERRY-HILL,
Twickenham, Middlesex.

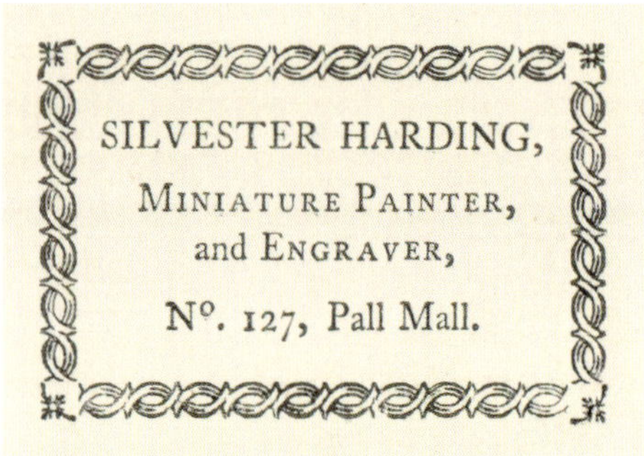

SILVESTER HARDING,
MINIATURE PAINTER,
and ENGRAVER,
N°. 127, Pall Mall.

THOMAS KIRGATE,
Late Printer at
STRAWBERRY-HILL,
Print and Bookseller,
N° 127, Pall-Mall.

56
Address card, "*Thomas Kirgate, Printer, Strawberry Hill,*" 1790s (before 1797).
Letterpress, 4.8 x 6.2 cm

57
Address card, "*Silvester Harding, Miniature Painter, and Engraver,*" ca. 1797.
Letterpress, 4.5 x 6.3 cm

58
Address card, "*Thomas Kirgate, Late Printer at Strawberry-Hill,*" ca. 1797.
Letterpress, 4.8 x 6.2 cm

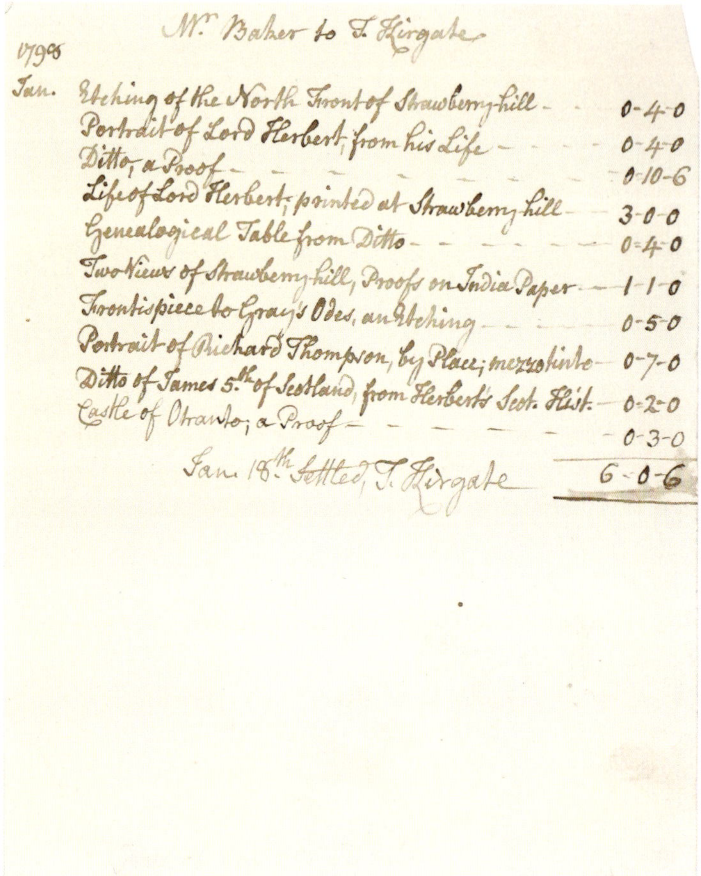

Mr. Baker to T. Kirgate

1798

Jan.	Etching of the North Front of Strawberry hill	0-4-0
	Portrait of Lord Herbert, from his Life	0-4-0
	Ditto, a Proof	0-10-6
	Life of Lord Herbert, printed at Strawberry hill	3-0-0
	Genealogical Table from Ditto	0-4-0
	Two Views of Strawberry hill, Proofs on India Paper	1-1-0
	Frontispiece to Gray's Odes, an Etching	0-5-0
	Portrait of Richard Thompson, by Place; mezzotinto	0-7-0
	Ditto of James 5th of Scotland, from Herbert's Scot. Hist.	0-2-0
	Castle of Otranto; a Proof	0-3-0
	Jan. 18th Settled, T. Kirgate	6-0-6

59
Invoice, Mr. Baker to T. Kirgate, Jan. 1798.
Pen and ink, 20.8 x 16.1 cm

Sir

George Harding had not finished the Drawing of the Duc de Nivernois till this morning, nor could I get Sir Kenelm Digby before yesterday, or I should have sent sooner. The late and present Sales afford nothing but what is extravagantly dear. A Guinea is a low price now for a scarce Print—upwards of *Forty* was given for one of a Duke of Norfolk, t'other day at Richardson's. Five, Ten, nay Twenty Guineas, is readily given for a Print, that has no other value but scarcity Illustrators of Granger, and other books, care less for Money than Portraits; the Rage for which increases—'tis a Madness which I hope will not reach Brixton [Bedford's address]. The few Ladies you want for p. 16 & 17, I do not think you will ever get; I bid upwards of a Guinea for one, the Duchess of Somerset, but she scorned such a paltry offer. Tis lucky, Sir, you want but such a few Prints to complete your Book; had you just begun to collect for it, you would soon quit the pursuit. I despair of finishing my Strawberry which I have now in hand.

I know not when I shall have the Pleasure of seeing you, Sir, for I can get nothing to bring; and am so infirm, I cannot walk far. But I shall not forget to search for the few you still want. As soon as I can see Bawtree [the engraver, friend of Kirgate and brother-in-law of Silvester Harding], I will get him to make a Drawing of the Bed in the Holbein Room, for I despair of getting Carter to do it Nor is Geo. Harding very willing to make any more Drawings—he has had the one now sent upwards of a Fortnight in hand.

I am, Sir,
your humble servt.
T. Kirgate
Sunday Night,
April 17th.[168]

This letter gives a clear impression of the compulsions of the Grangerizer, and of the spiraling prices that the fashion for extra-illustration had created. A sense of what was involved in "completing your Book" is evident in a two-page list prepared by Kirgate setting out desiderata of ninety-three portraits and other works of art in Walpole's collection, listing them by the page number from the *Description* in which they were mentioned; the list is annotated by him in red ink with comments such as "A foreign Print, 8vo.scarce," "Engraving by Harding," and, of three portraits in the Gallery, "I cannot say positively if there are Prints of these, the Pictures hanging too high to examine."[169] The letter also indicates the body of artists (Carter, George Perfect Harding, Bawtree) potentially available to provide Strawberry Hill images; Carter's extra-illustrated copy of the *Description* at the Huntington Library has a list of his sketches and drawings in the book, with their cost, part of which consists of a "List sent to Mr. Bull" of sixteen drawings with the prices, ranging from eight guineas for a view of the Tribune to five shillings for a bronze vase. As for the Hardings, in 1803 Silvester Harding published *A Series of Portraits, to Illustrate the Earl of Orford's Catalogue, of Royal and Noble Authors,* an eight-page pamphlet listing the authors mentioned and

60
George Perfect Harding, *Bed and Cardinal Wolsey's hat in the Holbein Chamber*, ca. 1800. Pen and ink and watercolor, 12 x 15 cm. Laid into Bawtree's extra-illustrated copy of the *Description* of Strawberry Hill

giving the numbers of the pages on which they were featured, so the collector had an aide-memoire of the prints he was searching for; Kirgate's copy has a tick against each name.[170] The activities of the Hardings, however, were not limited to Strawberry Hill; from 1795 to 1809, Silvester Harding, initially with his brother (who subsequently became librarian to Queen Charlotte), arranged the publication of *The Biographical Mirrour Comprising a Series of Ancient and Modern English Portraits, of Eminent and Distinguished Persons, from Original Pictures and Drawings.* This ultimately consisted of three quarto volumes, each with fifty portraits and supporting text; a number of the portraits (though only a small minority) were copied from originals at Strawberry Hill.

George Perfect Harding was also well known for his drawings and prints of historical figures, a good number of these from Walpole's collection; and his small, precise portraits are a recurrent feature of the extra-illustrated Strawberry Hill material put together around 1800 and in the first decades of the nineteenth century. He contributed substantially to William Bawtree's and to Kirgate's extra-illustrated copies of the *Description* and, for example, to Charles Bedford's extra-illustrated volume of "Poems by the Earl of Orford and Others printed at Strawberry Hill."[171] An indication of his industriousness is that the Lewis Walpole Library has well over one hundred Strawberry Hill-related drawings and watercolors by him. Some are of antiquities from Walpole's collection, some are views of the house and grounds, some are portraits of Walpole and his friends, and some are of specific items from the collection, such as the view of the "bed of purple cloth lined with white sattin" from the Holbein Chamber with Cardinal Wolsey's hat hanging next to it (Fig. 60); this was the watercolor that Kirgate had wanted Harding to reproduce for Charles Bedford.[172] His ongoing connection with the house is underscored by a picture he was to submit for exhibition at the Royal Academy in 1839 (the year in which he was elected a Fellow of the Society of Antiquaries), a view of the library at Strawberry Hill.

In 1801 Kirgate devised a raffle for his own benefit: seventy-three chances at two guineas a chance to acquire a copy of the *Description* of Strawberry Hill with at least three hundred drawings and prints, which he had spent fourteen years compiling (Fig. 61). This copy, now untraced, is distinct from his two other extra-illustrated copies now at the Lewis Walpole Library. Pending the raffle, the book could be inspected at Harding's bookshop at 127 Pall Mall. But there is a sense in which Kirgate's accumulation of images of Strawberry Hill was something more than a source of income. He had a collection of drawings and prints of Strawberry Hill pasted into an elephant folio: watercolors by the Hardings, including watercolors of the exterior and interior of the Printing House and the view from Kirgate's chamber (see Figs. 54, 64, and 80), the engravings of the antiquities bought by Walpole from Conyers Middleton, and miscellaneous prints. But these are by no means all images that he could arrange to have reproduced for collectors eager to grangerize their copies of the *Description;* they would in many cases no doubt have preferred more picturesque and general views. Instead, there are less readily sellable items, prints of the fireplaces and of antiquities as well as portraits, and the vignettes used by the Press, which would suggest that the collection was Kirgate's personal record. He even compiled a manuscript list of "Prints & Drawings as arranged in Lord Orford's book," which in fact consists of two separate lists, one of the prints and drawings in Walpole's extra-illustrated copy of the *Description,* the other of the prints and drawings in his own.[173] Despite the miserable legacy and the unwanted encounter with the harsh world beyond the Printing House door, despite his complaint about the extravagant cost of prints, and alongside and in addition to his trading in material from the Press, Kirgate was drawn into the Walpolian project of celebrating and recording Strawberry Hill. And quite clearly, the man whose sale catalogue contained not only numerous drawings of Strawberry Hill and innumerable prints, but also curiosities such as "A parcel various, containing a Curious Lock, a Roman Lamp, &c.," "Curious Carvings on Bones, presented by Lord Orford to Mr. Cole of Ely," and "A curious Carving in Ivory, and a Nutmeg Grater," had learned much from his master.[174]

There are two images of Kirgate from these years. The first is a pencil drawing of his head and shoulders in profile, with a dignified expression (Fig. 62; Catalogue 3.9). Beneath it is the inscription "Mr. Kirgate late Printer at Strawberry Hill for 30 Years. Drawn by T. Gosden from an Original Drawing," and beneath that is the larger vignette of Strawberry Hill. Drawing, inscription, and print are joined together by simple framing in pen and ink, surmounted by a festoon. The composition lies on the front endpaper of Thomas Gosden's bound manuscript "List of Pieces Printed at

61
"For the Benefit of T. Kirgate, Late Printer at Strawberry-Hill. To be Raffled for, a finely illustrated book, entitled, A Description of the Villa of Horace, Earl of Orford, at Strawberry-Hill, Twickenham," 1801.
Letterpress, 33.5 x 20.8 cm

For the BENEFIT *of* T. KIRGATE,

Late Printer at Strawberry-Hill.

TO BE RAFFLED FOR,

A FINELY ILLUSTRATED BOOK, ENTITLED,

A Description of the Villa

OF

HORACE, EARL OF ORFORD,

AT STRAWBERRY-HILL, TWICKENHAM.

EACH Leaf of the above Book is inlaid on a large Sheet of fine Dutch Paper; the Pages bordered with a double Line of red Ink; the Title printed in three Colours, and ornamented with a Drawing of a Foliage of Strawberries. The whole Book adorned with upwards of ~~Two~~ three Hundred Drawings and Prints, illustrative of the external and internal Parts of that delightful Seat, its Furniture, Pictures, Sculptures, Antiquities, &c. &c. Several of the Prints are scarce and valuable, and most of the Drawings were made on Purpose for the Book, by the Friends of T. KIRGATE, who has been Fourteen Years in forming the Collection.

IT is proposed to dispose of the above Book by Raffle, or some other Mode of Hazard, most agreeable to the Subscribers; to consist of Seventy-five Chances, at Two Guineas each; to be decided as soon as the Subscription is completed, and to be under the Direction of Ten or more of the Subscribers.

THE Book to be seen, and Subscriptions received, at No. 127, Pall-Mall, near Carlton-House.

Strawberry Hill," which is a copy of George Baker's *Catalogue of Books, Poems, Tracts, and Small Detached Pieces, Printed at the Press at Strawberry-Hill* (which was printed in 1810 in an edition of only twenty copies), so the drawing is probably of about 1810. Thomas Gosden (1780–1843) was a bookdealer, bookbinder, and sportsman who had worked in Piccadilly, Covent Garden, and St. Martin's Lane.[175]

The second portrait of Kirgate is a watercolor dated 1810, the year of his death, and fittingly is signed by George Perfect Harding (Fig. 63; Catalogue 3.10). It is of a man showing his years, a head and shoulders, looking to the left. It is in an oval painted frame, beneath which is the inscription "Thomas Kirgate Many years Printer to Horace Walpole Earl of Orford at Strawberry Hill. died 11 June 1810 aged 76." The expression is less confident than in the Gosden drawing, perhaps more hesitant, conceivably even resigned.

This is not to say that Kirgate softened in his account of his treatment by Walpole. He was visited at the end of his life by Dibdin, who was frustrated by his lack of interest in printing:

> I remember visiting Kirgate, who lived at the east end of Pall-Mall, not long before his decease. I was pretty earnest in my enquiries, but he was utterly destitute of all typographical enthusiasm; not a spark of *Paul Manutius, Oporinus,* or *Froben,* seems to have ever animated his bosom. He wore a brown curled wig, talked sparingly, but bitterly of Walpole's unkind treatment of him; and was civil to excess.[176]

One can imagine that the encounter between the energetic, eager, enthusing bibliophile and the tired, aging, long-retired printer would have brought little satisfaction to either of them.

As for the Printing House, there are three undated images that should be mentioned here.[177] First, Harding's interior view (Fig. 54) is one of a pair, the other being a watercolor of similar dimensions showing the exterior of the Printing House from the northeast (Fig. 64; Catalogue 1.17). It was presumably, like the interior view, painted at the very end of Kirgate's tenure, though the garden is not unkempt, and the building does not appear neglected, suggesting the late 1790s as a possible date. It does show detailing to the front garden fence different from that of the other images—Edwards showed close boarding, whereas the Birrell print after Harding (Fig. 41) showed closely placed posts and rails—but more significantly it shows what is probably a back door at the rear of the eastern side wall, leading out of the press room; otherwise it is the now familiar view, though with minor variants, such as the chamber window breaking the line of the cornice. The second image (Fig. 65; Catalogue 1.20) is similar to this, an etching by George Perfect Harding after Bawtree from the same vantage point, though it shows more trees to the left, and the chamber window here merely butts up to the lower rail of the cornice (which is now shown as being constructed of

62 (Cat. 3.9)
Thomas Gosden, *"Mr. Kirgate, late Printer at Strawberry Hill for 30 Years. Drawn by T. Gosden from an Original Drawing,"* with festoon above and large Strawberry Hill vignette below, ca. 1800–10. Pencil, 7.2 x 6.5 cm (portrait). Laid into Thomas Gosden's holograph "List of Pieces Printed at Strawberry Hill, taken from Baker's List"

brick). The back door in the eastern wall is loosely sketched in, and the fencing of the garden is consistent with the Birrell print after Harding. Third, there is a very simple amateur watercolor, probably from the first decade of the nineteenth century (Fig. 66; Catalogue 1.21). It is also from the same position, showing the building backed by trees, with the ivy on the facade between the composition room and chamber windows as seen in the Birrell print clearly visible. The picture has been cut around the outline of the trees, and the wall shielding the building from the Teddington road has been ignored. This cutout image of house and trees, with its heightened coloring (the mellow brick of the Edwards image is here rendered as bright red), and its rather primitive suggestion of the building's brickwork and fenced front garden, has something of the naive charm of a child's pop-up book illustration. Together with Gosden's drawing of Kirgate (Fig. 62), it is tipped into Gosden's manuscript copy of George Baker's Strawberry Hill *Catalogue,* and it is probably chronologically the last image we have of the Printing House in the original layout that Kirgate had known.

63 (Cat. 3.10)
George Perfect Harding, *"Thomas Kirgate Many years Printer to Horace Walpole Earl of Orford at Strawberry Hill. died June 11, 1810 aged 76,"* 1810.
Watercolor, 12.4 x 10 cm (image)

64 (Cat. 1.17)
Attributed to George Perfect Harding, *Exterior of the Printing House,* ca. 1790s. Pen and ink and watercolor, 18.5 x 23 cm. Laid into Kirgate's elephant folio collection of Strawberry Hill images. The Lord Waldegrave of North Hill

65 (Cat. 1.20)
George Perfect Harding, after William Bawtree, *"Printing-house at Strawberry-hill,"* ca. 1790s. Etching, 8.5 x 11.6 cm (image)

66 (Cat. 1.21)
Artist unknown, *"Printing Office at Strawberry-Hill,"* ca. 1810. Pen and ink and watercolor, 7 x 10 cm (max.). Laid into Thomas Gosden's holograph "List of Pieces Printed at Strawberry Hill, taken from Baker's List"

The clue to the next chapter in the building's history is provided by a footnote in Martin's *Bibliographical Catalogue of Books Privately Printed* (1834):

> The printing-office, on the death of lord Orford, was converted by Mrs. Damer into her modelling-room. Kirgate was considered by his noble employer as the only honest printer he ever had. For this, and some other information, the Editor is indebted to the kindness of Miss Berry, the accomplished editor of the Works of the Earl of Orford, in five volumes, 4to., published in 1798.[178]

Martin's appeal to the authority of Mary Berry, equally intimate friend of Walpole and of Damer, suggests that the conversion of the Printing House to a sculptor's studio is likely to be correct. We know that Damer worked at her sculpture at Strawberry Hill as well as at her house in Upper Brook Street, Mayfair, as Berry wrote in 1799 that "Mrs. Damer chips away at her marble one half of the morning, and trots about the grounds the other half in all weathers, and is much the better for this variety of exercise."[179] Percy Noble, in his biography of Damer, provides additional information about the sociability of her sculpting activity:

> Another theatrical friend of Mrs. Damer's was Mrs. Siddons, who was often a guest at Strawberry Hill, spending many hours in the studio, where, under the tuition of Mrs. Damer, she practised modelling... She, anyhow, owed much to Mrs. Damer, in whose studio she, with other fair dames arrayed in mob-caps and aprons, wielded the malet and chisel, kneading wax and clay with their white hands. Mrs. Siddons mentions that whenever she was with Mrs. Damer they indulged in their passion for sculpture, and adds that on one occasion, when she was on a visit to Mrs. Damer at Strawberry Hill, she had the honour of meeting Louis Philippe, afterwards king of France, and the Prince Regent.[180]

The style in which this information is presented may be anecdotal and unreferenced, but there is ample evidence of Sarah Siddons's friendship with Damer. And for all the period coyness of the image of Siddons as one of the fair dames with white hands, it seems likely that the Printing House may have enjoyed a brief Indian summer as the studio of the artist whom Walpole had compared with breathtaking assurance to Praxiteles, and as a studio may have entertained the most famous tragic actress of her generation.[181]

Damer gave up her life tenancy of Strawberry Hill in 1811, and the estate passed to Walpole's great-niece Elizabeth Laura, Lady Waldegrave (1760–1816). The decline of the Printing House over the following decade is summarized by Dibdin in an extended footnote to the *Bibliographical Decameron:*

> As Walpole wrote much and incessantly for his beloved PRESS, I was anxious to visit THE SPOT from which had issued unquestionably very many elegant and interesting publications: but what had been a receptacle only for *Long Primer,* and *Short Primer,* and *Pica,* and

67 (Cat. 1.22)
Artist unknown, *"The Printing Office erected by Horace Walpole at Strawberry Hill,"* early 1820s.
Pen and ink and wash, 17.8 x 25 cm (image)

> *English, &c.* types, had been long ago converted into a *brew-house,* or *wash-house,* at this moment I forget which.[182]

Just how the Printing House was allowed to decay is evidenced by what is at first sight its most perplexing depiction (Fig. 67; Catalogue 1.22). It is a gray wash drawing, one of a series of eleven, all on similar paper and apparently by the same hand, one of them watermarked "J Whatman Turkey Mills 1821," another similarly watermarked but dated 1822, so it may well date from 1821 or soon after.[183] But for the title added in pencil, "The Printing Office erected by Horace Walpole at Strawberry Hill," one would not think at first glance that it bore any relation to Kirgate's Printing House. It shows a square brick box, with a door in one corner, and with a large louvered opening above and to the left of it. A sash window in the side wall to the left has been bricked up. The roof on the left of the building is hipped, but at the right end has no such hip, the unseen righthand side wall abutting the far side of the chimney.

68 (Cat. 2.5)
Detail of *Twickenham Enclosure Award Map: Plan F*, 1818. Pen and ink and wash on parchment, 67 x 84 cm. Bound into the Enclosure Award of 1818. Local Studies Library, London Borough of Richmond upon Thames

Yet this is in fact the surviving eastern half of the Printing House, reduced to a brew house or wash house (for which the double height of the former press room would have been an asset, as it would also have been when it was used as a sculptor's studio), or similar menial use. The original Printing House consisted of the press room to the east, and to the west the composing room, with Kirgate's chamber above: a conjectural reconstruction of this is provided in Plan I on page 119. What Fig. 67 shows is the former press room, its sash window extended and replaced by a louvered window, the second sash in its eastern, side wall bricked up, and the front door of the building shorn of its projecting porch. The door, the chimney, and the bricked-up sash window in the side wall are all in the same positions as shown in the Edwards group of views (Figs. 24–27), Carter's pencil drawing (Fig. 45), and the Harding watercolor (Fig. 64); but the righthand, two-storied side of the building (the composing room and chamber above) has been demolished. The proportions of the sketch are not exact, and it seems slightly to foreshorten the width of the former press room between the door and the eastern flank wall with its bricked-up window, but the relationship of door, chimney, and window, and the shape and pitch of the surviving section of the tiled roof, are clear.[184] The brick cornice reflects that suggested in the Birrell print (Fig. 41) and the etching by George Perfect Harding after Bawtree (Fig. 65). Behind the building, on the left, is Walpole's new ice house of 1795. All trace of Kirgate's cottage front garden has disappeared. How the truncated building might have been laid out is suggested in Plan II.

Confirmation that part of the building had been demolished is provided by the Twickenham Enclosure Award map of 1818 (Fig. 68; Catalogue 2.5). This clearly shows—shaded pink—Walpole's house and the Offices and, set back from the farmyard, the distinctive flask-like shape of the new ice house. Opposite it is the greenhouse (marked "E" on the Carter plan in Walpole's copy of the *Description* [Fig. 11]); and whereas the Carter plan shows the Printing House (marked "F") as a rectangular building extending virtually up to the Teddington road, the Enclosure Award map shows a smaller building, square in plan, set back from the road—that is, occupying the eastern half of the site of the former Printing House.

There is no documentation of the use or fate of what remained of the Printing House during the next few decades. In 1842, the great Strawberry Hill sale scattered Walpole's carefully preserved treasures across the country and beyond; the sale was spectacular, the publicity surrounding it extensive, but there is no mention of the Printing House. The little publicized and subsequently aborted auction of "Excellent Furniture throughout the Mansion and Offices of the Earl of Waldegrave, at Strawberry Hill" to be held on 12 and 13 July of that year, largely concerned with the service apartments and outhouses, gives no mention in its catalogue of any lots specifically identified with the Printing House, but it does conclude with ten lots that might conceivably relate to the former Printing House, being equipment from the brew house, ending with a final humble lot of "a deal table, on horse legs."[185]

The only picture of the Printing House that we have from later in the nineteenth century is a conjectural reconstruction (Fig. 69; Catalogue 1.23); it is a pen-and-ink drawing on card, labeled on the rear "Lewis Buddy III Feb. 1946 (E.C. Buddy) The SH Printing Office." Lewis Buddy (1872–1941) was a printer, type designer, and book collector and member of the Rowfant Club, with a particular interest in Horace Walpole—he used two pictorial bookplates, both featuring Strawberry Hill. The book in which Catalogue items 1.21 and 3.9 (Figs. 66 and 62) are laid belonged to him, and W.S. Lewis acquired some material from his family after his death, though the main sale of his collection appears to have taken place in 1910, when the Anderson Auction Company offered 659 lots, of which seventy-seven were Walpole or Strawberry Hill Press items. Buddy also in 1901 printed, at his Kirgate Press in Canton, Pennsylvania, three hundred copies of *Horace Walpole and the Strawberry Hill Press 1757–1789*, illustrated with items from his collection (including this sketch of the Printing House).[186] The drawing shows a compacted version of the original building, copying its fenestration, but clad inaccurately in weatherboarding, and with proportions if anything rather closer to the boxlike structure shown in the wash drawing of the 1820s (Fig. 67); it is an imaginative remodeling of a building already changed almost beyond recognition.

The transformation of Strawberry Hill in the mid-nineteenth century is centered on the extraordinary personality of Frances, Lady Waldegrave (1821–1879; Fig. 70). She was the daughter of the popular opera singer John Braham (1777–1856), and the story has often been told of her spectacular rise through society, marrying first

69 (Cat. 1.23)
Artist unknown,
The Strawberry Hill Printing Office, 19th century. Pen and ink, 10.4 x 18.8 cm

John Waldegrave (1814–1840), elder but illegitimate and alcoholic son of the sixth Earl; swiftly followed by his brother George, the seventh Earl (1816–1846); then after a decent interval of twelve months following his death from cirrhosis of the liver, the "short staid widower of sixty-one," George Granville Harcourt (1785–1861); and finally, just over a year after Harcourt's death, Chichester Fortescue, a cabinet member under Gladstone and later Lord Carlingford (1823–1898) — representing respectively a foothold in society, a title, money, and high politics.[187] Lady Waldegrave inherited Strawberry Hill after the death of her second husband, and before the death of her third she had started restoring the house, buying back items from Walpole's collections when opportunity arose, and entertaining there, so transforming the house into the venue for the Whig and Liberal party "Friday to Monday" parties for which she became famous.

After initial alterations to Walpole's house in 1856–58, Lady Waldegrave proceeded in 1860–62 to build the suite of entertaining rooms, now known as the Waldegrave wing, linking Walpole's house and his Offices. She had the route of the Teddington road moved away from the house to make space for her new building, and this also enabled her to create a garden and entrance sweep between the road and the north front of the house, and to build a new U-shaped stable yard of London stock brick opening onto the road, over and to the southwest of Walpole's farmyard. This can be seen in the Ordnance Survey map of 1863 (Fig. 71; Catalogue 2.7).[188] Following Lady Waldegrave's death, the bereaved Lord Carlingford began to dismantle the estate. He initially offered for sale as building plots Strawberry Hill Road and Waldegrave Park Gardens (Fig. 72; Catalogue 2.8) and then, in 1883, sold the balance of the estate

70
Anonymous artist, after James Rannie Swinton, *Frances, Countess of Waldegrave,* 1850. Lithograph with added body color, 25 x 20.4 cm (visible). The Hon. Nino Strachey

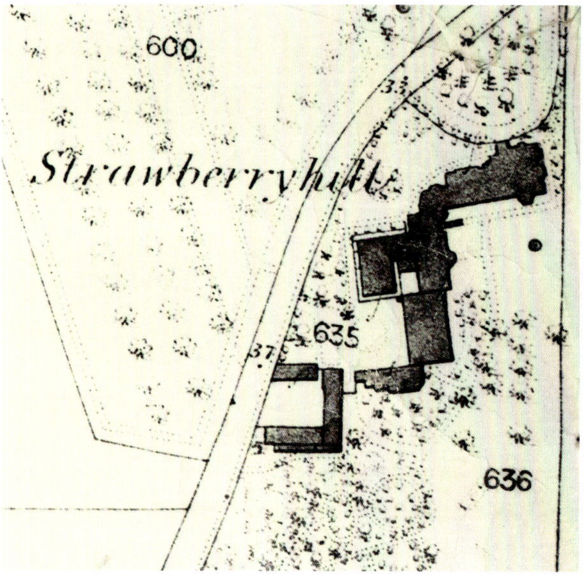

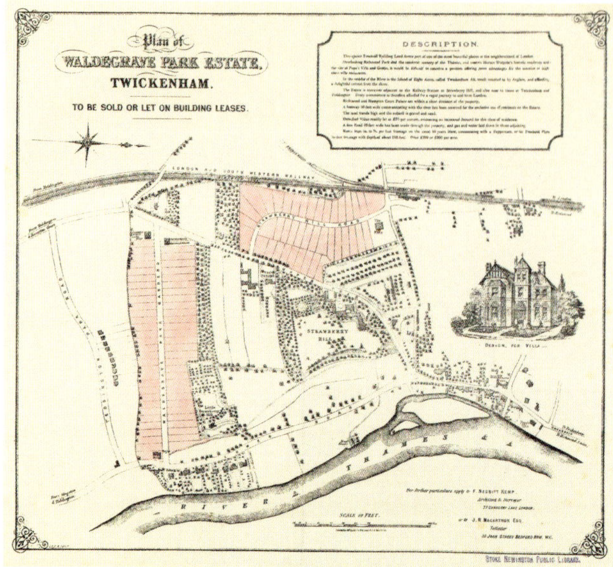

71 (Cat. 2.7)
Strawberry Hill and its stable block, Ordnance Survey 25" Map of Middlesex, Sheet xx:15 (detail), surveyed in 1863. Zincograph, 69 x 101 cm. Local Studies Library, London Borough of Richmond upon Thames

to Baron de Stern (1815–1887). We have the sale particulars with a plan of the estate both for that sale (Fig. 73; Catalogue 2.9) and for an apparently abortive auction sale in 1888 following Baron de Stern's death (see Fig. 82, showing a detail with the house and stable block; Catalogue 2.10). The plans show the same configuration of Lady Waldegrave's estate buildings; the former Printing House has been lost in her ambitious building scheme. In 1923 the estate passed to the Catholic Education Council on behalf of the Vincentian Community, from whose use of the property to train teachers evolved what is today St Mary's University College. The University buildings now spread far to the south and west of Walpole's and Lady Waldegrave's work, but her solidly constructed stable block survives.

And there the tale of the decline, decay, and disappearance of Walpole's Printing House might have rested, save for the survival among the extraordinary holdings of the Lewis Walpole Library of a set of photographs. These date from about 1900, and consist of views from two different perspectives of a small, unpretentious building (Figs. 74 and 75; Catalogue 1.24). In one of them (not illustrated, but a less focused version of Fig. 74), there is a man with a coat and rather crumpled hat at its front door; on the back of the card on which that photograph is mounted, W.S. Lewis has written "From Lewis Buddy III Taken at SH c. 1900." A comparison of these photographs with the 1820s wash drawing (Fig. 67) reveals that the Printing House in its truncated form survived Lady Waldegrave's building campaign, being reconstructed and incorporated into her stable block. A large rectangular patch of infill brickwork has been inserted where the wash drawing showed the louvered opening, but the brick

72 (Cat. 2.8) *opposite* *"Plan of Waldegrave Park Estate, Twickenham. To be Sold or Let on Building Leases,"* ca. 1880. Hand-colored lithograph, 21 x 26.3 cm

73 (Cat. 2.9) *"The Strawberry Hill Estate Twickenham. For Sale by Messrs. Ventom, Bull & Cooper,"* 1883. Hand-colored lithograph, 88 x 59 cm

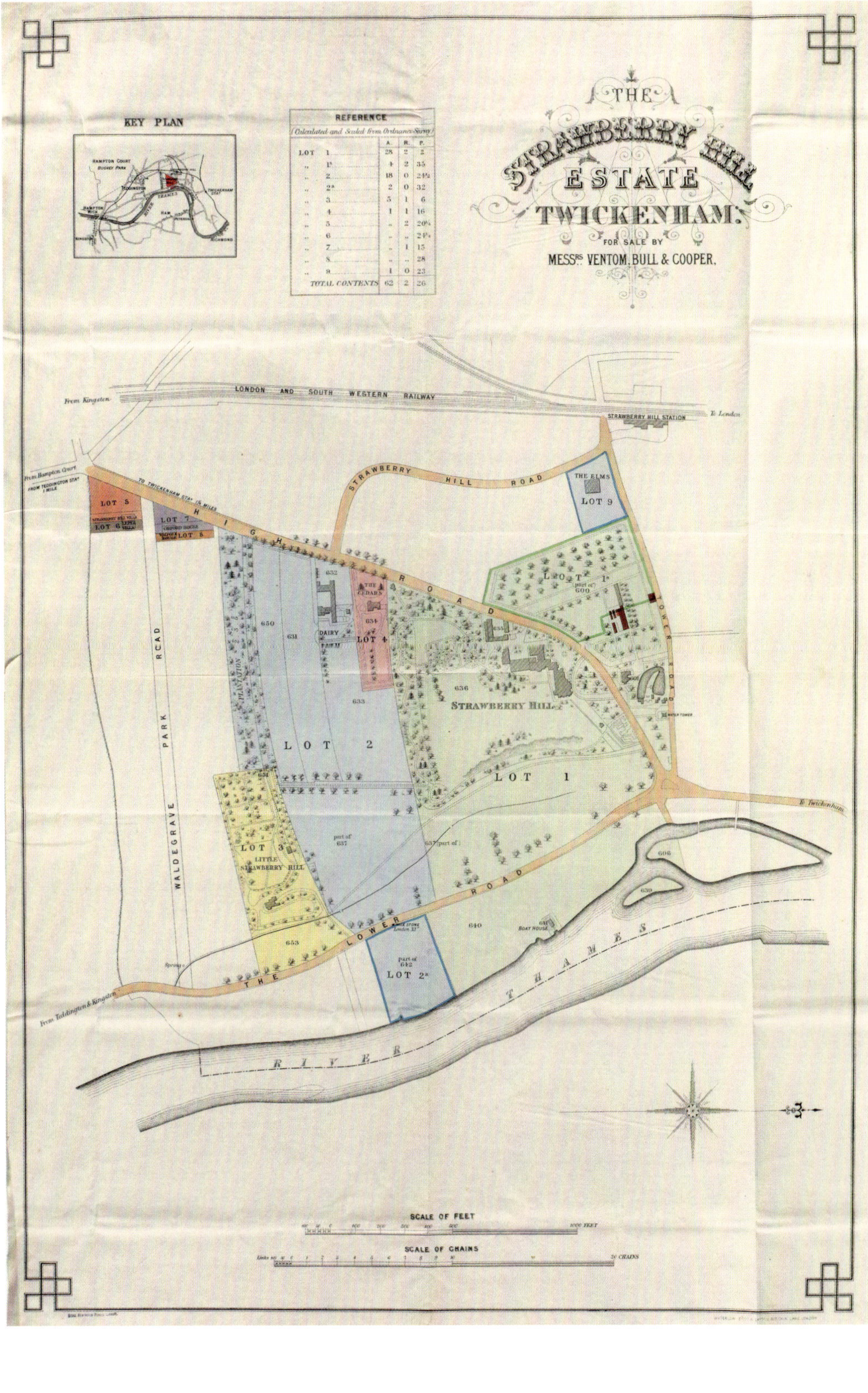

74 (Cat. 1.24.1)
The Printing House at Strawberry Hill, view from the north, ca. 1900.
Gelatin silver print, 17.8 x 12.8 cm

75 (Cat. 1.24.5)
The Printing House at Strawberry Hill, view from the northeast, ca. 1900.
Gelatin silver print, 12.8 x 17.7 cm

stepped cornice of the building in the wash drawing is identical to the cornice clearly shown in the photographs. What has changed most materially is that the building has been reconverted to domestic use. The front door has been moved to the east, nearer the center of the remaining structure, and the roof has been renewed and re-covered in slate, with four pitches rising to an apex on which are new chimney pots. Two new windows have been created, one at ground-floor level to the left of the door, the other a distinctive narrow window immediately above the door.[189] Fig. 75, showing the building from the same angle as the wash drawing, reveals that the east, side wall has been rebuilt with two windows at the upper-floor level, breaking into the brick cornice, suggesting that the interior has been reconfigured as two floors throughout. The side elevation also shows two blind window reveals at ground-floor level; to the rear, the building is joined to Lady Waldegrave's stable block. A projected configuration of the building as so converted is shown as Plan III, with the northern end of the side walls of Lady Waldegrave's building visible, and with steep stairs lit by the window above the door separating a pair of small rooms on the ground floor and a pair of rooms above.

Returning to the Ordnance Survey map of 1863 (Fig. 71) and the estate sale plan of 1883 (Fig. 73), we can identify the building as the square block at the extreme northeast corner of the stable yard, projecting slightly further west than the north/south run of the Waldegrave building behind it that forms the east wing of the stable yard. Comparing this plan to the 1818 Enclosure Award map (Fig. 68) and—making allowance for the reduction of the building from rectangle to square—to the Carter plan (Fig. 11), we can see that its position is unchanged; the rear, southern wall of the Printing House in all its transformations remained parallel to but just south of the southern wall of Walpole's Offices. All that has changed between the Enclosure map and the 1880s sale plans is the position of the Teddington road and the presence of Lady Waldegrave's adjoining stable buildings.

And so, if you come today to Strawberry Hill, what will you see? If you walk past the north, entrance front of Walpole's house, leaving his Round Tower behind you, and bear left past the side of the Waldegrave wing, you are turning south, with the rear of the Waldegrave wing on your left, and the Teddington road (now named Waldegrave Road) on your right. Beyond the Waldegrave wing, the open courtyard area that had lain immediately to the west of Walpole's Offices, seen in the Ordnance Survey map of 1863 and the 1880s sale plans, has been filled with a building designed by the architectural firm Pugin and Pugin for the Vincentian fathers in the 1920s, when the property was a Catholic teacher training college. The area across which you are walking is now used for parking by University staff, and at its south end, facing you, is the north block of Lady Waldegrave's stable courtyard (Fig. 76). The former stable courtyard itself, originally open to the Teddington road, is now choked with

76
The Printing House (left) and former stable building (right), Strawberry Hill, 2010. Photo: Christine Jarvis, St Mary's University College

an amalgam of later and temporary structures that serve the University's building maintenance needs, but all three solidly constructed stable blocks remain. Looking at the north side of the northern block, you see on the right her gray stock brickwork; then a gap, as shown in the Ordnance Survey and the 1880s plans; and then, on the left, forming the north end of her eastern stable block, a small, unprepossessing, square building (Fig. 77; Plan IV).

It is not identical in appearance to the photographs of 1900; its brick has been covered in gray render, and it has been given a very utilitarian projecting front porch with a modern door. The window to the left of the front door now has an arched head, and on the eastern, side elevation, the northern ground-floor blank window has been opened, while the southern one has disappeared beneath a single-story brick corridor linking it to the Pugin and Pugin building (Fig. 78). But, beneath all these indignities, a comparison of Figs. 74 and 75 with 77 and 78 shows that the remains of Walpole's Printing House survive. The narrow window above the door in the photographs of 1900 (lighting the stairs to what is now an upper-floor flat) remains, as do the building's proportions, its slate roof, its relationship to Lady Waldegrave's stable blocks, and Walpole's stepped, brick cornice (now masked by render). Inside, no original fittings survive. It now consists at ground-floor level of two small offices facing north, divided by a wall supporting the steep stairs that rise directly from the front door to the flat above; the chimney has disappeared, and the chimney pots visible in the photographs of 1900 have been capped. At the rear of the ground floor is a corridor, separating the two offices from the Waldegrave stable blocks; the doorway at the east end of this

77
The Printing House, view from the north, 2010.
Photo: Christine Jarvis, St Mary's University College

78
The Printing House, view from the northeast, 2010.
Photo: Christine Jarvis, St Mary's University College

corridor, opening onto the modern corridor that connects to the Pugin and Pugin building, approximates the position of the rear doorway seen in Harding's watercolor and etching (Figs. 64 and 65), while the doorway at its west end occupies the same space as the doorway between the press room and the composing room in Carter's interior views (Figs. 46 and 47). The stairs would have been inserted when the building was reconverted, presumably as part of the 1860s building campaign that created the stable blocks. The living accommodation to which they give access extends across the entirety of the building; the airspace of the former press room is now occupied by a kitchen to the west of the stairs, and a bedroom and bathroom to the east; the flat also extends south, over the first floor of the east wing of Lady Waldegrave's stable block. The spaces are small, confirming the suspicion that—making all due allowance for the successive transformations that the building has experienced—the Carter view may have somewhat exaggerated the size and airiness of the Printing House, and in particular its depth.

Yet the surprising and previously unrecorded fact is that the press room of Walpole's Printing House, the space in which the books and pamphlets and ephemera of the Press were printed, still exists: divided, reroofed, refaced, refenestrated, and wholly unrecognizable to anyone who looked merely at the Edwards or Harding images. But with the benefit of the wash drawing of the 1820s and the photographs of ca. 1900, its transformation is revealed. And so, against all odds, uncelebrated and unexpected, a significant part of the Printing House of Walpole's Strawberry Hill Press survives, a survival in its way quite as surprising as that of Strawberry Hill itself.

Notes

1 "Short Notes of the Life of Horatio Walpole, Youngest Son of Sir Robert Walpole, Earl of Orford, and of Catherine Shorter, His First Wife," in *The Yale Edition of Horace Walpole's Correspondence,* ed. W.S. Lewis et al. (New Haven and London: Yale University Press, 1937–83), 13: 28.

2 *Correspondence* 35: 98–99.

3 Gray wrote to James Brown on 25 July 1757: "you must know (what you will like no more, than I do; yet it was not in my power any how to avoid it) Mr W:, who has set up a printing-press in his own house at Twickenham, earnestly desired, that he might print it for Dodsley, & as there is but one hand employ'd, you must think it will take up some time to dispatch 2000 Copies" (in *Correspondence of Thomas Gray,* ed. Paget Toynbee and Leonard Whibley [Oxford: Clarendon Press, 1935], 2: 508–9). One suspects that Gray's sense of frustration may not have been caused merely by the resultant delay in printing the edition.

4 Walpole to Sir Horace Mann, 4 August 1757, *Correspondence* 21: 120.

5 Walpole to Sir Horace Mann, 12 June 1753, *Correspondence* 20: 381.

6 Walpole to Richard Bentley, 3 November 1754, *Correspondence* 35: 186. For an account of the hall and staircase and of the interiors of the house generally, see Michael Snodin, "Going to Strawberry Hill," in *Horace Walpole's Strawberry Hill,* ed. Michael Snodin (New Haven and London: Lewis Walpole Library, Yale Center for British Art, and Victoria and Albert Museum, in association with Yale University Press, 2009), 14–57, esp. 31 and 34.

7 Paget Toynbee, ed., *Journal of the Printing-Office at Strawberry Hill: Now First Printed from the MS. of Horace Walpole* (London: Constable and Company Limited; and New York: Houghton Mifflin Company, 1923), 4–5.

8 Antony Griffiths, "Sir William Musgrave and British Biography," *British Library Journal* 18, no. 2 (Autumn 1992): 171–89, esp. 180–1.

9 George E. Haggerty, "Strawberry Hill: Friendship and Taste," in *Horace Walpole's Strawberry Hill,* 74–85, esp. 75 and 77.

10 For the Strawberry Committee see George E. Haggerty, "The Strawberry Committee," and John Riely, "Horace Walpole, the 'Strawberry Committee,' and the 'Committee of Taste,'" in *Horace Walpole's Strawberry Hill,* 80–81 and 349, respectively.

11 *Correspondence* 28: 195.

12 John Heneage Jesse, *George Selwyn and His Contemporaries: with Memoirs and Notes* (London: R. Bentley, 1843), 1: 310. Also discussed in W.S. Lewis, *Horace Walpole,* The A.W. Mellon Lectures in the Fine Arts 9 (London: Rupert Hart-Davis, 1961), 43.

13 For the Guthrie pamphlet see *Correspondence* 38: 437–38.

14 Morris R. Brownell, *The Prime Minister of Taste: A Portrait of Horace Walpole* (New Haven and London: Yale University Press, 2001), 2–5.

15 Walpole to Sir Horace Mann, 7 September 1781, *Correspondence* 25: 184; Walpole to William Mason, 4 August 1782, *Correspondence* 29: 272; and Walpole to Sir Horace Mann, 9 May 1779, *Correspondence* 24: 475.

16 R.W. Ketton-Cremer, *Horace Walpole: A Biography,* 3rd ed. (London: Methuen & Co. Ltd., 1964), 21.

17 Cynthia Roman, "The Art of Lady Diana Beauclerk: Horace Walpole and Female Genius," in *Horace Walpole's Strawberry Hill,* 155.

18 Walpole to Sir David Dalrymple, 8 November 1767, *Correspondence* 15: 112.

19 William Beloe, *The Sexagenarian; or, the Recollections of a Literary Life* (London: F.C. and J. Rivington, 1817), 1: 276–81, at 277–78. For a more balanced account of Walpole as host see John Pinkerton, *Walpoliana* (London, 1799), 1: xli–xliii.

20 Lucy Peltz, "A Friendly Gathering: The Social Politics of Presentation Books and Their Extra-Illustration in Horace Walpole's Circle," *Journal of the History of Collections* 19, no. 1 (2007): 37 and 42. See also her "Engraved Portrait Heads and the Rise of Extra-Illustration: The Eton Correspondence of the Revd James Granger and Richard Bull, 1769–1774," *The Walpole Society* 66 (2004), 1–161; Robert R. Wark, "The Gentle Pastime of Illustration," *Huntington Library Quarterly* 56, no. 2 (Spring 1993): 154–58; and Ruth Mack, "Paper Castle, Paper Collection:

Walpole's Extra-Illustrated Copy of the *Description of... Strawberry Hill*," in *Horace Walpole's Strawberry Hill*, 106–15.

21 Walpole to William Cole, 16 June 1781, *Correspondence* 2: 273.

22 Walpole to Lord Strafford, 13 June 1781, *Correspondence* 35: 359; and *Correspondence* 33: 362, n. 26, for the quotation from W.P. Courtney. The summary of Storer's character in Lord Glenbervie's *Diaries* (ed. Francis Bickley [1928], 2: 349) is quoted in *Correspondence* 11: 114, n. 12.

23 See the account of the marriage portrait of Bull in Ellen G. D'Oench, *The Conversation Piece: Arthur Devis and His Contemporaries* (New Haven: Yale Center for British Art, 1980), cat. 19. See also *Polite Society by Arthur Devis, 1712–1787: Portraits of the English Country Gentleman and His Family* (Preston: Harris Museum and Art Gallery, 1983), cat. 18, p. 48, for a discussion of the relationship between the Bull portrait and Devis's very similar portrait of Robert Dashwood and his wife.

24 Letter dated 2 February 1792, laid into John D. Fry's extra-illustrated copy of the *Description* of Strawberry Hill, compiled 1894; Lewis Walpole Library, Yale University, 33 30 copy 30. Bull's extra-illustrated copy of the *Description*, now at the Lewis Walpole Library, contains a genealogy of the Walpole family "to Explain the Portraits and Coats of Arms at Strawberry Hill, Anno 1776," together with Carter's drawings of the head of Henry III from Barnwell church and of the bust of Caligula (illustrated in *Correspondence* 2: facing p. 275; and 22: facing p. 522). The letters between Walpole and Bull are in volumes 41 and 42 of the *Correspondence*. Bull also sent for Walpole's perusal his extra-illustrated copies of the Strawberry Hill Press *Grammont* (Walpole to Richard Bull, 20 December 1781, *Correspondence* 41: 467), the *Royal and Noble Authors* (Walpole to Lady Ossory, 25 December 1782, *Correspondence* 33: 376), and Lucan's *Pharsalia* (Walpole to Richard Bull, 14 July 1784, *Correspondence* 42: 107–8).

25 See note 7.

26 John Bruce and the Rev. Thomas Thomason Perowne, eds., *Correspondence of Matthew Parker, D. D. Archbishop of Canterbury, Comprising Letters Written by and to Him from A.D. 1535, to His Death, A.D. 1575* (Cambridge: Cambridge University Press for the Parker Society, 1853), 42: 426. For a brief summary of other eighteenth-century private presses, see *Horace Walpole Printer: A Paper Read by Edward Percival Merritt at a Meeting of the Club of Odd Volumes XVII April MCMVII* (Boston: privately printed, 1907), 7–11. For an extended account, see David Chambers, *The Private Press in Britain, 1701–1900* (Private Libraries Association, forthcoming).

27 Horace Walpole, *Memoirs of the Reign of King George III*, ed. Derek Jarrett (New Haven and London: Yale University Press, 2000), 1: 116–17; Walpole to William Mason, 4 June 1782, *Correspondence* 29: 253.

28 Farmer subsequently passed on the proof sheet so that in due course it reached the authorities. See John Sainsbury, *John Wilkes: The Lives of a Libertine* (Aldershot: Ashgate, 2006), 153; and for this incident and generally as an account of Wilkes's press, see James Mosley and David Chambers, "The Printing Office of John Wilkes in Great George Street, 1763," in *The Private Library*, 6th ser., vol. 1, no. 4 (Winter 2008): 148–88, esp. 164, 158 (which reproduces the page of *The North Briton*, no. 46, with the inaccurate attribution to Strawberry Hill), and 159.

29 I am grateful to David Chambers for information on the Grange Press.

30 A.T. Hazen, *A Bibliography of the Strawberry Hill Press with a Record of the Prices at Which Copies Have Been Sold* (New Haven: Yale University Press, 1942).

31 *Journal of the Printing-Office*, 10.

32 Walpole to Sir David Dalrymple, 15 October 1759, *Correspondence* 15: 58.

33 Walpole to John Chute, 12 July 1757, *Correspondence* 35: 98–99.

34 Walpole to Hannah More, 2 July 1789, *Correspondence* 31: 305.

35 Fanny Burney, Madame D'Arblay, *Memoirs of Doctor Burney* (London, 1832), 3: 65–66.

36 Walpole to Edward Jerningham, 30 July 1789, *Correspondence* 42: 253.

37 Walpole to the Earl of Harcourt, 10 August 1773, *Correspondence* 35: 459.

38 *Anecdotes of Painting in England; with Some Account of the Principal Artists; and Incidental Notes on Other Arts; Collected by the Late Mr. George Vertue; and Now Digested and Published from His Original MSS. By Mr. Horace Walpole* (Strawberry Hill, 1762–71). Frederick W. Hilles and Philip B. Daghlian, eds., *Anecdotes of Painting in England [1760–1795]... Volume the Fifth and Last* (New Haven: Yale University Press, 1937). For later editions of and supplements to the *Anecdotes*, see Hazen, *Bibliography of the Strawberry Hill Press*, 62–63 and 65–66.

39 Lionel Cust, "George Vertue's Note-books and Manuscripts Relating to the History of Art in England," *Third Annual Volume of the Walpole Society, 1913–14* (Oxford: Oxford University Press, 1914), 122; Lawrence Lipking, *The Ordering of the Arts in Eighteenth-Century England* (Princeton: Princeton University Press, 1970), 163. For the most recent discussion of the *Anecdotes* in the context of their European progenitors, see Nigel Llewellyn, "*The Anecdotes of Painting* and Continental European Art History," in *Horace Walpole's Strawberry Hill*, 136–53.

40 Brownell, *The Prime Minister of Taste,* 93; and Horace Walpole, *Memoirs of King George II,* ed. John Brooke (New Haven and London: Yale University Press, 1985), 2: 2.

41 W.S. Lewis, *Horace Walpole's Library* (Cambridge: Cambridge University Press, 1958), 11; reprinted in Allen T. Hazen, *A Catalogue of Horace Walpole's Library* (Oxford: Oxford University Press; and New Haven: Yale University Press, 1969), 1: liv.

42 For the complicated history of the printing of the *Description* of Strawberry Hill and the evolution of its text, see Stephen Clarke, "'Lord God! Jesus! What a House!': Describing and Visiting Strawberry Hill," *Journal for Eighteenth-Century Studies* 33, no. 3 (September 2010): 357–80.

43 As explained in Hazen, *Bibliography of the Strawberry Hill Press,* 123, which adds that a copy of the list is bound into Thomas Crofton Croker's expanded and extra-illustrated copy of the Catalogue of the Strawberry Hill Sale of 1842, now at the Lewis Walpole Library, Yale University, 485 842 c76 vi copy 5; a further copy is reproduced in Nicolas Barker, ed., *Horace Walpole's Description of the Villa at Strawberry-Hill: A Facsimile of the Copy Extra-Illustrated for Charles Bedford in the Collection of Lord Waldegrave of North Hill* (London: The Roxburghe Club, 2010), at the rear of the facsimile section.

44 George Robins, *A Catalogue of the Classic Contents of Strawberry Hill Collected by Horace Walpole* (London, 1842).

45 Barker, ed., *Horace Walpole's Description of the Villa at Strawberry-Hill,* 33–45; Mack, "Paper Castle, Paper Collection," 108–12. Barker provides (p. 149) a helpful listing of the seventeen known extra-illustrated copies of the *Description* (including closely related illustrated items such as Kirgate's elephant folio collection), followed by a detailed finding list of the illustrations in them.

46 *Horace Walpole's Strawberry Hill,* cat. 101, p. 297. The bookcase initials are among the Strachie papers at the Somerset Heritage Centre, Taunton (ref: dd/sh/57/209), and consist of a set of printed initials N to T and numbers 1 to 5. They would appear to relate to the book room in the Offices built by Walpole in 1790, which W.S. Lewis describes as being divided into presses or bays N to V (*Horace Walpole's Library,* 19–20), and may well represent further unrecorded items from the Press.

47 H.J. Jackson, *Marginalia: Readers Writing in Books* (New Haven and London: Yale University Press, 2001), 204, 232–33.

48 Walpole to Bishop Lyttelton, 10 July 1763, *Correspondence* 40: 287.

49 Walpole to Lord Ossory, 23 June 1771, *Correspondence* 32: 45–46.

50 Walpole to William Cole, 18 February 1773, *Correspondence* 1: 300.

51 Walpole to Sir David Dalrymple, 23 February 1764, *Correspondence* 15: 98.

52 Walpole to Thomas Walpole the Younger, 8 April 1786, *Correspondence* 36: 238. Walpole complained to Lady Ossory of having been cheated by the publisher John Bell over the fourth and final volume of the *Anecdotes of Painting* but did not specify the grounds of his allegation (Walpole to Lady Ossory, 6 February 1787, *Correspondence* 34: 35–36).

53 George Montagu to Walpole, 30 August 1764, *Correspondence* 10: 134; Sir David Dalrymple to Walpole, 26 September 1764, *Correspondence* 15: 103.

54 Chambers, *The Private Press in Britain, 1701–1900* (forthcoming).

55 In the Northwick Park collection of Loose Pieces, Lewis Walpole Library, Yale University, 336J.

56 Walpole to William Mason, 22 September 1783, *Correspondence* 29: 312.

57 Thomas Gray to Walpole, 10 August 1757, *Correspondence* 14: 99; and Henry Seymour Conway to Walpole, 15 October 1758, *Correspondence* 37: 578.

58 Walpole to William Mason, 15 May 1773, *Correspondence* 28: 88.

59 Walpole to William Mason, 25 September 1771, *Correspondence* 28: 24.

60 Walpole to Lady Ossory, 16 July 1789, *Correspondence* 34: 52.

61 Walpole to Barbara Cecilia Seton, 17 December 1790, *Correspondence* 11: 166.

62 Walpole to Richard Bull, 10 January 1787, *Correspondence* 42: 185.

63 This copy is at the Lewis Walpole Library, Yale University, 33 14 copy 5.

64 Hilles and Daghlian, eds., *Anecdotes of Painting in England,* 202, and *Correspondence* 1: 287, n. 5. For a poignant account of Gulston's rape of the retiring Cole's collection of portrait prints, see William Cole to Walpole, 20 November 1772, *Correspondence* 1: 287–88.

65 Walpole to Lady Ossory, 15 September 1787, *Correspondence* 33: 573–77.

66 Pinkerton, *Walpoliana* 2: 163–64. See also Walpole's letter of 15 September 1787 to Lady Ossory, *Correspondence* 33: 575–76, for his account of how the copy of the 1774 edition of the *Description* of Strawberry Hill that he had given to William Cole entered the trade after Cole's death.

67 John Martin, *A Bibliographical Catalogue of Books Privately Printed; Including Those of the Bannatyne, Maitland and Roxburghe Clubs, and of the Private Presses at Darlington, Auchinleck, Lee Priory, Newcastle, Middle Hill, and*

Strawberry Hill (London, 1834), 2: 490. The two Cruden letters are in a private collection. The first of them is in fact addressed to S. Harding, although Silvester Harding had died ten years earlier. This is probably simply an error, perhaps because Cruden had been used to dealing with Silvester Harding in building up his collection.

68 *Bibliotheca Walpoliana: Being the Third Supplement to a General Catalogue for 1842 of the Most Interesting, Curious, and Rare Collection of Books, on Sale in the World. Selected from the Elegant Library of Horace Walpole, Earl of Orford, Recently Dispersed at Strawberry Hill...By Thomas Thorpe* (1842); and *Strawberry Hill: A Catalogue of a Singularly Choice Collection of Rare and Valuable Books, in the Various Departments of Literature...Principally from the Collection of Horace Walpole...By W. Strong, Bristol* (1843).

69 Walpole to Lady Ossory, 16 July 1789, *Correspondence* 34: 53.

70 *Correspondence* 13: 28; *Journal of the Printing-Office,* 3; and Walpole to Lord Strafford, 13 September 1759 and 30 October 1759, *Correspondence* 35: 296 and 298.

71 Michael Snodin, "Strawberry Hill: Building and Site Part Two: The Site," *Architectural History* 38 (1995): 123–25.

72 Paget Toynbee, ed., *Strawberry Hill Accounts: A Record of Expenditure in Building Furnishing, &c. Kept by Mr Horace Walpole from 1747 to 1795* (Oxford: Clarendon Press, 1927), 1.

73 Walpole to Sir Horace Mann, 4 August 1757, *Correspondence* 21: 120; *Strawberry Hill Accounts,* 7.

74 Walpole to Sir Horace Mann, 4 August 1757, *Correspondence* 21: 120–21.

75 Walpole to Henry Zouch, 15 March 1759, *Correspondence* 16: 27. According to the *Journal of the Printing-Office,* 8, Robinson went away on 5 March 1759.

76 *Journal of the Printing-Office,* 8.

77 Walpole to Sir David Dalrymple, 22 May 1759, *Correspondence* 15: 55–56.

78 Walpole to Sir David Dalrymple, 19 June 1759, *Correspondence* 15: 57.

79 *Journal of the Printing-Office,* 8; Walpole to Sir David Dalrymple, 15 October 1759, *Correspondence* 15: 59.

80 *Journal of the Printing-Office,* 9; Walpole to Lord Strafford, 30 October 1759, *Correspondence* 35: 298. See also Walpole's previous letter to Lord Strafford of 13 September 1759, in which he states that "I have begun to build a new printing-house, that the old one may make room for the Gallery and Round Tower" (*Correspondence* 35: 296).

81 *Journal of the Printing-Office,* 9; *Strawberry Hill Accounts,* 8.

82 W.S. Lewis, in "The Genesis of Strawberry Hill," *Metropolitan Museum Studies* 5, pt. 1 (1934): 85, points out that the plan must postdate the erection of the Offices in 1790, and predate November 1791, when Walpole became Lord Orford.

83 Walpole's offices were built by a Mr. Chapman, builder of Twickenham, and not by James Wyatt, as stated by Howard Colvin in *A Biographical Dictionary of British Architects, 1540–1840,* 4th ed. (New Haven and London: Yale University Press, 2008), 1185, following W.S. Lewis in "The Genesis of Strawberry Hill," 82, in turn following the rather more qualified comments of Toynbee, *Strawberry Hill Accounts,* 175–76; see Clarke, "'Lord God! Jesus! What a House!,'" 380, n. 41.

84 The very rough outline of the ice house is shown in Walpole's gouty sketch of the farmyard (Fig. 23; Cat. 1.2). The *Strawberry Hill Accounts* records an expenditure of £210-15-10¾ for the building of an ice house and cottage in 1769.

85 This is one of a set of eleven wash drawings of the house and estate, of which another example is Fig. 67 (Cat. 1.22), discussed below.

86 Walpole to Mary Berry, 8 September 1795, *Correspondence* 12: 163. The new ice house in its altered position is also shown in A. Barham's "Plan of Strawberry Hill" of 1797 (which curiously does not show the Printing House), reproduced in *Strawberry Hill Accounts* following p. 192.

87 Walpole to George Montagu, 17 May 1763, *Correspondence* 10: 71.

88 Walpole to George Montagu, 18 June 1764, *Correspondence* 10: 127.

89 Walpole to George Montagu, 11 May 1769, *Correspondence* 10: 278–79. The verses are reprinted in the *Journal of the Printing-Office,* 14–16.

90 Lewis Walpole Library, Yale University, 336G.

91 George Montagu to Walpole, 23 June 1763, *Correspondence* 10: 84.

92 Walpole to George Montagu, 25 August 1757, *Correspondence* 9: 215–16.

93 Walpole to Lady Ossory, 19 October 1788, *Correspondence* 34: 26–27.

94 Walpole to George Montagu, 23 December 1761, *Correspondence* 9: 411.

95 Walpole to William Cole, 20 May 1762, *Correspondence* 1: 12. Walpole's manuscript account of Farmer's disappearance and his melancholy parting from his friends and neighbors survives and is printed in the *Journal of the Printing-Office,* 81–82.

96 *Journal of the Printing-Office,* 10, 84–86. One earlier entry (p. 79) shows Pratt being paid at a guinea a week.

97 Walpole to George Selwyn, 28 September 1763, *Correspondence* 30: 171–72.

98 *Journal of the Printing-Office,* 85. See also William Bathoe to Walpole, 27 August 1763, *Correspondence* 40: 290; and 10: 102, n. 1.

99 *Correspondence* 30: 171, n. 3.

100 *Journal of the Printing-Office,* 13.

101 In regard to Kirgate's year of birth, Kirgate annotated a proof state of Edwards's print of 1784 (Fig. 36; Cat. 3.1) with the note "Aetat 49"; and the portrait of 1810 by George Perfect Harding (Fig. 63; Cat. 3.10) is inscribed by Harding that Kirgate was 76 when he died on 11 June that year.

102 *Journal of the Printing-Office,* 17.

103 *Correspondence* 41: 237; Walpole to William Mason, 8 December 1773, *Correspondence* 28: 121; see also Walpole to William Cole, 3 and 15 January 1779, *Correspondence* 2: 134 and 138, acknowledging that his printer Kirgate was now his secretary.

104 Walpole to Lady Ossory, 25 December 1782, *Correspondence* 33: 378.

105 *Correspondence* 11: 166; 12: 31; 42: 414; 15: 265; and 11: 111, 370, and 306. For Kirgate showing Strawberry Hill to visitors, see the "Book of Visitors" in *Correspondence* 12: 217–52, at 224, 226, 231, and 232 ("3 artists by Kirgate"); see also 12: 141 and 167–68. Kirgate's copy of the Long Gallery picture hang is at the Guildhall Library, MS 01628.

106 Clarke, "'Lord God! Jesus! What a House!,'" 373–76.

107 Walpole to Hannah More, 2 July 1789, *Correspondence* 31: 306; Walpole to Lady Ossory, 18 July 1789, *Correspondence* 34: 53.

108 Walpole to Mary Berry, 20 October 1791, *Correspondence* 11: 370; Walpole to Lady Ossory, 26 December 1791, *Correspondence* 34: 135.

109 Walpole to Lady Ossory, 27 January 1786, *Correspondence* 33: 512.

110 See Hazen, *Bibliography of the Strawberry Hill Press,* 234–48.

111 The will is printed in *Correspondence* 30: 344–77.

112 Kenneth Garlick and Angus MacIntyre, eds., *The Diary of Joseph Farington* (New Haven and London: Yale University Press, 1979), 3: 785–86.

113 Pinkerton, *Walpoliana* 1: xxiv, n.; Lady Theresa Lewis, ed., *Extracts of the Journals and Correspondence of Miss Berry from the Year 1783 to 1852* (London, 1865), 2: 65. See also Henry Lemoine, *Typographical Antiquities: History, Origin, and Progress, of the Art of Printing…also a Particular and Complete History of the Walpolean Press, Established at Strawberry Hill; with an Accurate List of Every Publication Issued Therefrom, and the Exact Number Printed Thereof…* (London, 1797), 94, in which Lemoine comments that "Mr. Thomas Kirgate, the printer and conductor of all his Lordship's literary labours for more than 40 years, only stands in his patron's will on the footing of a menial servant, by a legacy of ONLY 100 pounds."

114 W.S. Lewis, *The Forlorn Printer: Being Notes on Horace Walpole's Alleged Neglect of Thomas Kirgate* (Farmington: privately printed, 1931).

115 *Correspondence* 17: xlii, n. 4; and 43: 232, respectively referring to and quoting from the footnote on page xvi of volume 1 of Mrs. Paget Toynbee, ed., *The Letters of Horace Walpole* (Oxford: Clarendon Press, 1903–5). Note 4 of *Correspondence* 17: xlii, also adds that Mrs. Thomas's letter was first printed in *Willis's Current Notes* (August 1854), 62.

116 Hazen, *Bibliography of the Strawberry Hill Press,* esp. 12–14. Hazen analyzes the reprints in the individual entries for Books and Detached Pieces under the heading "States and Variants."

117 Ibid., 26–29 and 74–76; A.T. Hazen and J.P. Kirby, *A Bibliography of the Strawberry Hill Press* (1942; reprint, with a supplement, Folkestone and London: Dawsons of Pall Mall, 1973), ix.

118 Barker, *Horace Walpole's Description of the Villa at Strawberry-Hill,* 99.

119 Lot 421 on the second day, recorded as withdrawn in the priced copy of the catalogue, Lewis Walpole Library, Yale University, 486 10.

120 Walpole to the Hon. Thomas Walpole, 29 December 1780, *Correspondence* 36: 189 and n. 7; and William Mason to Walpole, 18 January 1783, *Correspondence* 29: 281. As to Mason's reliability, there is a painful contrast between Walpole's pleading with him in 1774 and again in 1783 to provide copy for the Press (*Correspondence* 28: 162; and 29: 310) and Mason's self-righteous but untruthful declaration to his patron Lord Harcourt in 1790, after they were both distanced from Walpole, that "For *myself,* I am thankful I never had an offer of the use of that press even in the most cordial times" (*Correspondence* 29: 363).

121 This is not to say that Walpole was incapable of disinterested acts of charity; Barker, in *Horace Walpole's Description of the Villa at Strawberry-Hill,* notes instances of Walpole's anonymous acts of charity as recorded in five letters to his deputy Grosvenor Bedford (107, n. 1); *Correspondence* 40: 270 and 345; and 42: 454–55.

122 Lewis, *The Forlorn Printer,* 12; Nathaniel Hillier to Walpole, 19 October 1780, *Correspondence* 16: 95–97. Hillier was a merchant, antiquary, and collector.

123 British Museum, Anderdon, *Collectanea Biographica* 55: 34. Anderdon's accompanying note records that Kirgate was "for many years Printer & companion" to Walpole, "for whose long & faithful Services he left him by his Will the miserable sum of £100."

124 Hazen, *Bibliography of the Strawberry Hill Press,* 248; the note is on Lewis Walpole Library copy 336c.

125 Ibid., 154–58.

126 See *Horace Walpole's Strawberry Hill,* 45 (reproducing the watercolor, Fig. 65) and cat. 131, p. 306; Walpole to William Cole, 16 June 1781, *Correspondence* 2: 274. Edwards's watercolors engraved for the plates in the 1784 edition of the *Description* include the engraved frontispiece, the Entrance, the Stairs, the Library, and the Gallery.

127 Edward Edwards, *Anecdotes of Painters Who Have Resided in or Been Born in England; with Critical Remarks on their Productions* (London, 1808), vi. For the Beauclerk Cabinet, see Susan Odell Walker, "An Ebony Cabinet Ornamented with Drawings by Lady Diana Beauclerk," in *Horace Walpole's Strawberry Hill,* 160–61.

128 Walpole to Mary Berry, 6 November 1795, *Correspondence* 12: 175.

129 The Rev. T. F. Dibdin, *The Bibliographical Decameron; or, Ten Days Pleasant Discourse upon Illuminated Manuscripts, and Subjects Connected with Early Engraving, Typography, and Bibliography* (London, 1817), 3, 450.

130 Ibid., 356–63 (Hafod) and 453–55 (Auchinleck); there is also a woodcut of the Auchinleck printing house in Martin, *A Bibliographical Catalogue of Books Privately Printed* 2: 347.

131 Martin, *A Bibliographical Catalogue of Books Privately Printed* 2: 441–58; A. N. L. Munby, *Portrait of an Obsession: The Life of Sir Thomas Phillipps, the World's Greatest Book Collector, Adapted by Nicolas Barker from the Five Volumes of* Phillipps Studies (London: Constable; and New York: G. P. Putnam's Sons, 1967), 32; also 51–57.

132 Walpole to Mary Berry, 27 September 1794, *Correspondence* 12: 111.

133 Martin, *A Bibliographical Catalogue of Books Privately Printed* 2: 379. Martin provides accounts that are still of interest of the Grange, Auchinleck, Lee Priory, Middle Hill, and Strawberry Hill presses. See also *The Autobiography, Times, Opinions, and Contemporaries of Sir Egerton Brydges...* (London, 1834), 2: 191–94, and Dibdin, *The Bibliographical Decameron* 3: 455–68, for the Lee Priory Press.

134 Hazen, *A Catalogue of Horace Walpole's Library* 3: 244–46; the copy is now in the collection of the Earl of Rosebery.

135 G. Barrett of 289 High Holborn had another connection with Strawberry Hill, in that in 1796 he also published a plate of the painting at Strawberry Hill that Walpole believed to be the "Marriage of Henry the 7th," which had originally appeared as engraved by Grignion in the first volume of the *Anecdotes of Painting.*

136 Lewis Walpole Library, Yale University, folio 33 30 copy 11, f. 240.

137 The print of Kirgate by Edwards as finished by Hall appears as the forty-seventh of fifty-one etched plates in *A Collection of Views and Studies after Nature with Other Subjects Designed and Etched by Edward Edwards, Associate and Teacher of Perspective in the Royal Academy* (London, 1790). For another example of Walpole's proprietorial attitude regarding the control of images and the position of artists, see his letter of 16 June 1781 to William Cole, *Correspondence* 2: 274–75, in which he records his disapproval of the conduct of the artist William Watts, who had complained in print of being ejected from Chiswick House for taking a view of the house for publication without leave.

138 We do have a portrait by Harding of Ann Branscom, Walpole's housekeeper, who on occasion showed the house to visitors, but it almost definitely postdates Walpole's death. It is laid into Kirgate's extra-illustrated copy of the *Description* of Strawberry Hill, Lewis Walpole Library, Yale University, 33 30 copy 6.

139 So described by Walpole in the *Description,* 2.

140 For David Edward Cronin, who trained in London and Paris and during the Civil War provided illustrations for *Harper's Weekly* under the pseudonym Seth Eyland, see Robert R. Wark, "The Gentle Pastime of Illustration," *Huntington Library Quarterly* 56, no. 2 (Spring 1993): 162. A copy of *The Poems and Plays of Oliver Goldsmith,* ed. Austin Dobson (London, 1889) in the collection of the Huntington Library, San Marino, California, 109574, inscribed on the title page as "Hand illustrated by David Edward Cronin, New York, 1889," has more than 200 illustrations by him. Thanks to Lesley Baier for making this connection.

141 For a general account of Carter and his passionate engagement with the Gothic (though not dealing in any detail with his activities at Strawberry Hill), see J. Mordaunt Crook, *John Carter and the Mind of the Gothic Revival* (London: The Society of Antiquaries of London, Occasional Paper no. 17, 1995).

142 The major Carter watercolors of Strawberry Hill are reproduced as Figs. 45, 48, 61, 72, 77, 81, 128, 245, and 260 of *Horace Walpole's Strawberry Hill.* Of the two *Otranto* watercolors, *The Entry of Prince Frederick into the Castle of Otranto* and *The Death of Matilda,* the first is reproduced in *Horace Walpole's Strawberry Hill,* Fig. 13, and both are reproduced in Crook, *John Carter and the Mind of the Gothic Revival,* 12–13.

143 Huntington Library, San Marino, California, 130368; Hazen, *Bibliography of the Strawberry Hill Press,* 262.

144 Stated in two similarly worded holograph notes at the Lewis Walpole Library, Yale University, one in the Portrait File, Kirgate folder, on a sheet beneath Barrett's portrait of Kirgate (Fig. 37; Cat. 3.2) and initialed W. U. [William Utterson?], the other formerly laid beneath a portrait of Kirgate but now separated from it, currently

in the Kirgate folder of the box of Strawberry Hill prints and photographs.

145 Herbert Davis and Harry Carter, eds., *Joseph Moxon, Mechanick Exercises on the Whole Art of Printing, 1683–4* (Oxford: Oxford University Press, 1958), 17.

146 Walpole to John Chute, 26 July 1757, *Correspondence* 35: 100; and Walpole to William Mason, 15 May 1773, *Correspondence* 28: 87–88.

147 Thomas Farmer to Walpole, 27 August 1759, *Correspondence* 40: 164.

148 *Journal of the Printing-Office,* 9; Walpole to Sir David Dalrymple, 15 October 1759, *Correspondence* 15: 59.

149 *Journal of the Printing-Office,* 20; 68, n. 2; and 88 (Appendix G); Walpole to the Duc de Nivernois, 1 February 1785, *Correspondence* 42: 135.

150 In the Northwick Park Collection of Loose Pieces, Lewis Walpole Library, 336J.

151 *Journal of the Printing-Office,* 95–96.

152 Hazen, *Bibliography of the Strawberry Hill Press,* 11, 91.

153 Walpole to Grosvenor Bedford, 9 November 1757, *Correspondence* 40: 117.

154 Hazen, *Bibliography of the Strawberry Hill Press,* 11.

155 Walpole to Sir David Dalrymple, 22 May 1759, *Correspondence* 15: 56.

156 Davis and Carter, eds., *Joseph Moxon, Mechanick Exercises,* 17.

157 Walpole to Lady Ossory, 18 July 1789, *Correspondence* 34: 53.

158 The smaller vignette was, however, used later in the printed title pages that Walpole provided to his octavo "Collection of the Most Remarkable Tracts Published in the Reign of King George the Third," and to his collections of Poems and of Dramatic Pieces published in the reign of King George the Third, all ca. 1768; and the larger vignette was used by Kirgate on some versions of the title page that Walpole provided for his "Collection of the Loose Pieces Printed at Strawberry Hill." See Hazen, *Bibliography of the Strawberry Hill Press,* 255–57, 264.

159 Hazen, *A Catalogue of Horace Walpole's Library* 1: xxv and plates IV and V; *Journal of the Printing-Office,* 2; 23, n. 2; and 77 (which shows that Grignion charged £2-12-6 for engraving the larger vignette and £2-2-0 for the smaller). Hazen corrects Toynbee by stating that the smaller vignette was designed by Müntz rather than Bentley.

160 The rectangular view of the house is also used on the title pages printed at the Press for Walpole's quarto "Collection of the Most Remarkable Tracts Published in the Reign of King George III," his "Portraits of the Reign of King George the Third," and his "Additions to the Collection of English Portraits." See Hazen, *Bibliography of the Strawberry Hill Press,* 258–60.

161 My thanks to John Gambell for this suggestion.

162 The Lewis Walpole Library has a number of pamphlets from the Press and two copies of the *Description,* all sewn in this manner, some using a single and some a double thread, sewn into very similar drab blue, laid-paper wrappers. Walpole's accounts, particularly with Thomas Farmer, do contain references to the purchase of "2 doz. of Skinns" (*Journal of the Printing-Office,* 79–80), but these would have been used for the tympan and perhaps the frisket on the common press, and also for covering the ink balls, not for binding (though the number of skins purchased in just two years is perhaps surprising).

163 For a general life of Anne Damer, see Percy Noble, *Anne Seymour Damer: A Woman of Art and Fashion, 1748–1828* (London: Kegan Paul, Trench, Trübner & Co. Ltd., 1908). For a contemporary assessment of her sculpture, see Allan Cunningham, *The Lives of the Most Eminent British Painters, Sculptors, and Architects* (London, 1830), 3: 247–73; and for a modern assessment, see Alison Yarrington, "The Female Pygmalion: Anne Seymour Damer, Allan Cunningham and the Writing of a Woman Sculptor's Life," *Sculpture Journal* 1 (1997): 32–44.

164 For the relevant passages of Walpole's will, see *Correspondence* 30: 374, 354.

165 *Correspondence* 30: 374.

166 Hazen, *Bibliography of the Strawberry Hill Press,* 277.

167 See also Lewis, *The Forlorn Printer,* 17. George Baker, in addition to compiling *A Catalogue of Books, Poems, Tracts, and Small Detached Pieces, Printed at the Press at Strawberry-Hill, Belonging to the Late Horace Walpole, Earl of Orford* (1810, twenty copies), was an antiquarian and the county historian of Northamptonshire. There is also at the Lewis Walpole Library, 335 F1, an invoice for £15-3-6 from Kirgate to the collector Francis Annesley for thirty-one Strawberry Hill items.

168 Kirgate MSS. Folder, Lewis Walpole, Yale University; this letter has been printed in Barker, *Horace Walpole's Description of the Villa at Strawberry-Hill,* 116.

169 The list is bound into Strawberry Hill Records, Lewis Walpole Library, Yale University, 486 20.

170 Lewis Walpole Library, Yale University, 33 16 copy 12.

171 Lewis Walpole Library, Yale University, 33 30 copies 4 and 6, and 335R. For George Perfect Harding, as for his father Silvester Harding, see the *Oxford Dictionary of National Biography* articles by F.M. O'Donoghue, revised respectively by Mervyn Cutten and J. Desmarais.

172 *A Description of the Villa of Mr. Horace Walpole...at Strawberry-Hill* (Strawberry Hill, 1784), 43.

173 Bound into Strawberry Hill Records, Lewis Walpole Library, Yale University, 486 20.

174 The items quoted from Kirgate's sale catalogue are lots 43, 48, and 67 on the tenth day.

175 There is in the Beinecke Rare Book and Manuscript Library, Yale University, Sheldon Collection JAX 099, an extra-illustrated copy of the *Description* on the title page of which is laid what appears to be the title (only) of a print, with the engraved signature "T. Kirgate, Strawberry-Hill" and, printed beneath it, "Engraved from the Original in the Possession of Mr. T. Gosden." The lettering is consistent with a steel engraving, and it may be the title of an untraced print of Kirgate, perhaps related to this drawing of him by Gosden.

176 Dibdin, *The Bibliographical Decameron* 3: 451, note. The two similarly worded holograph notes referred to at note 144 above detail his moves to Pall Mall and then to Vine Street, where he died in 1810, and refer to the sale of his books, prints, and curiosities by Messrs. King & Lochée on 3 December 1810 and ten following days.

177 There is also a fourth, "ghost" view, being an aquatint of the Cottage in the Garden seen from the rear, engraved by John Watts after J.C. Barrow, the image 11 x 12.2 cm, mistitled "The Printing House at Strawberry Hill." An example is laid into Anthony Storer's atlas folio of Strawberry Hill material titled "Aedis Walpoli" (Eton College Library, Storer Bequest). A copy of the same print in Kirgate's elephant folio collection of Strawberry Hill material replaces the engraved words "The Printing House" with the word "Cottage"—with a scroll at the end of the word to help fill the space. A copy of the "Cottage" version is reproduced in the facsimile section of Barker, *Horace Walpole's Description of the Villa at Strawberry-Hill.*

178 Martin, *A Bibliographical Catalogue of Books Privately Printed* 2: 489, note.

179 Lewis, *Extracts of the Journal and Correspondence of Miss Berry* 2: 103.

180 Noble, *Anne Seymour Damer,* 164–65.

181 For Walpole's inscription on Damer's terracotta model of a fishing eagle, comparing her to Praxiteles, see Walpole to Lady Ossory, 14 June 1787, *Correspondence* 33: 564 and facing plate. For Siddons's friendship with Damer, see Thomas Campbell, *Life of Mrs. Siddons* (London, 1834), 2: 301; and Shearer West, "The Public and Private Roles of Sarah Siddons," in *Sarah Siddons and Her Portraitists,* ed. Robyn Asleson (Los Angeles: The J. Paul Getty Museum, 1999), 28 and 39, n. 75.

182 Dibdin, *The Bibliographical Decameron* 3: 449–50, note.

183 Lewis Walpole Library, Yale University, Chewton Box, 49 36 78. These gray wash drawings are part of the same group as the gray wash drawing of the rear facade of the Offices (Fig. 13, there titled "The Stables at Strawberry Hill").

184 Although generally correct as to detail, the proportions of others of the group of gray wash drawings can be inaccurate.

185 Stephen Clarke, "The Strawberry Hill Sale of 1842: 'The Most Distinguished Gem that has ever Adorned the Annals of Auctions,'" in *Horace Walpole's Strawberry Hill,* 260–74, at 269–70. A copy of the very scarce sale catalogue of "Excellent Furniture" is bound into volume 4 of Thomas Mackinlay's extra-illustrated copy of George Robins's auction catalogue, *A Catalogue of the Classic Contents of Strawberry Hill,* Lewis Walpole Library, Yale University, 485 842 C76 IV copy 3.

186 Munson Aldrich Havens, *Horace Walpole and the Strawberry Hill Press, 1757–1789* (Canton, Pa.: Lewis Buddy 3rd The Kirgate Press, 1901), facing p. 48, beneath a reproduction of the portrait of Kirgate by Gosden that is Fig. 62 (Cat. 3.9). The 1909 auction by the Anderson Auction Company was their sale no. 779–1909.

187 Osbert Wyndham Hewett, *Strawberry Fair: A Biography of Frances, Countess Waldegrave, 1821–1879* (London: John Murray, 1956), 56; but note that the *Oxford Dictionary of National Biography* article on Lady Waldegrave states that at the time of their marriage, Harcourt was 62, not 61, and had a daughter five years older than Lady Waldegrave. For an entertaining account of Lady Waldegrave and her building work at Strawberry Hill, see J. Mordaunt Crook, "Strawberry Hill Revisited—III," *Country Life* 52, no. 3965 (21 June 1973): 1794–97.

188 The copy of the Ordnance Survey map illustrated (Fig. 71; Cat. 2.7) was based on the survey of 1863 and was first published in 1863/4, being the first map to show Lady Waldegrave's completed stable yard. It is, however, conceivable that the eastern range of Lady Waldegrave's stable block may date from the mid-1840s, although it appears to be of the same build as the northern and southern ranges of 1860–62. The map of the parish of Twickenham by William Thomas Warren of 1846 (Fig. 81; Cat. 2.6) shows a southern extension to the remains of the Printing House, essentially consistent in plan with the eastern range of Lady Waldegrave's stable block, but there is no record of any building works at Strawberry Hill after Lady Waldegrave returned from abroad with the seventh Earl, who died in September 1846.

189 Peter Inskip has pointed out that the window to the left of the door in the photographs of ca. 1900, with its star-shaped upper lights, is similar to the windows of the facade of Walpole's much extended and reconfigured but still surviving Cottage (the former Cottage in the Flower-Garden) facing Strawberry Hill across what is now Waldegrave Road.

Plans: The Changing Configuration of the Printing House

The Printing House of the Strawberry Hill Press
An Iconography

Notes to the Reader

Titles derived from inscriptions on the object are shown in quotation marks.

Measurements are in centimeters, height before width. Laid and wove papers have been distinguished where possible.

In the case of multiple copies of an item, emphasis has been placed on copies at the Lewis Walpole Library, Yale University, given the depth of its Walpole and Strawberry Hill holdings; copies are from that collection unless specified otherwise. References to the *Description* are to Walpole's *A Description of the Villa of Mr. Horace Walpole, Youngest Son of Sir Robert Walpole Earl of Orford at Strawberry-Hill near Twickenham, Middlesex. With an Inventory of the Furniture, Pictures, Curiosities, &c.* (Strawberry Hill, 1784). Note that Walpole's and Richard Bull's extra-illustrated copies of the *Description* have been digitized in their entirety by the Lewis Walpole Library and are available online at www.library.yale.edu/walpole/collections/digital_collection.html under call numbers 49 3582 and 33 30 copy 11, respectively.

Section One: Views of the Printing House

{ 1.1 ❧ fig. 9 }

"END VIEW OF STRAWBERRY HILL BEFORE THE GALLERY"
Probably George Perfect Harding (1779/80–1853), copy made ca. 1790s after a lost drawing of 1754
Pencil and watercolor on wove paper, 21 x 33 (image), 25 x 35 (including ruled edging)
Title inscribed in pencil in Kirgate's hand, below image
Laid into Kirgate's elephant folio collection of Strawberry Hill images
The Lord Waldegrave of North Hill

Bibliography: Michael Snodin, "Strawberry Hill: Building and Site. Part Two: The Site," *Architectural History* 38 (1995): 116–28, at 123–25.

{ 1.2 ❧ fig. 23 }

"VIEW OF FARMYARD & PRINTING HOUSE AT STRAWBERRY HILL."
Horace Walpole (1717–1797), probably early 1780s
Pen and ink on laid paper, 16.1 x 19.9
Title inscribed in pen and ink in Walpole's hand above lower margin of sheet
Annotated in pencil in Kirgate's hand below image: "Barn. Ice House. Printing House"
Inscribed on the verso in pen and ink in Kirgate's hand: "Sketch by Mr. Walpole (when his Hands were wrapt in Bootikins in a fit of the Gout) for a Print, before *that* by Edwards was made" [for the print after Edwards, see Catalogue 1.5]
The Lewis Walpole Library, Yale University

A pencil note below Kirgate's inscription—"See His letter No 1163"—refers to letter 1163 in Peter Cunningham, ed., *The Letters of Horace Walpole, Earl of Orford,* 9 vols. (1857–59), a letter of 2 December 1768 to Sir Horace Mann, the last paragraph of which discusses Walpole's, Mann's brother's, and Chute's gout and the efficacy of wearing bootikins.

{ 1.3 ❧ fig. 24 }

"PRINTING HOUSE AT STRAWBERRY HILL 1784"
Edward Edwards (1738–1806), 1784
Pen and ink and watercolor on wove paper, laid onto card with wash-line mount, 17.9 x 23.8 (image), 25.4 x 31.6 (mount)
Title inscribed in pen and ink on mount
The Lewis Walpole Library, Yale University

Reproduced: W.S. Lewis, *Collector's Progress* (New York: Alfred A. Knopf, 1951), facing p. 96.

The northeast view, with a barn to the left, the ice house just visible between it and the Printing House, and Kirgate walking in front of the Printing House, portfolio under his arm.

There is an early note in pencil at foot of mount: "The original drawing M.A. Rooker"; with a later note in W.S. Lewis's hand: "? E. Edwards"; the attribution to Michael Angelo Rooker (1743–1801) is of no authority, the aquatint (Catalogue 1.5) confirming that the watercolor is by Edwards.

See also Catalogue 1.4, 1.5, and 1.6.

{ 1.4 ❧ fig. 25 }

"PRINTING HOUSE AT STRAWBERRY HILL"
Edward Edwards (1738–1806), 1783
Etching and watercolor on laid paper, laid onto separate sheet bearing wash-line mount, 18 x 23.3 (image), 22.4 x 27.8 (mount)
Signed and dated in the plate, within image: "E.E. 1783"
Title inscribed in pen and ink on mount
Inscribed on verso: "The Seat of the Honle. H. Walpole, in the Parish of Twickenham"
The Lewis Walpole Library, Yale University

Bibliography: Michael Snodin, with the assistance of Cynthia Roman, ed., *Horace Walpole's Strawberry Hill* (New Haven and London: Lewis Walpole Library, Yale Center for British Art, and Victoria and Albert Museum, in association with Yale University Press, 2009), 323, cat. 199, fig. 350.

Exhibition History: *Horace Walpole's Strawberry Hill,* Yale Center for British Art, New Haven, and Victoria and Albert Museum, London, 2009–2010, no. 199.

The same view as Catalogue 1.3, 1.5, and 1.6.

{ 1.5 ❧ fig. 26 }

"FARM YARD AND PRINTING HOUSE AT STRAWBERRY HILL."
Edward Edwards (1738–1806) and Francis Jukes (1745–1812), after Edwards, 1784
Etching and aquatint on laid paper, 18.2 x 23.6 (image), 20.8 x 25.3 (plate)
Signed and dated in the plate, within image: "E.E. 1783"
Inscribed in the plate: "Drawn & Etch'd by E Edwards. A. Engrav'd in Aquatinta by F Jukes. / Farm Yard and Printing House at Strawberry Hill. / London Published as the Act directs Feby. 1st. 1784 by F. Jukes No. 1, Great Marylebone Street."
The Lewis Walpole Library, Yale University

Additional copies are laid into Walpole's manuscript "Journal of the Printing Office at Strawberry-hill near Twickenham in Middlesex," facing p. 16; Walpole's and Richard Bull's extra-illustrated copies of the *Description,* 49 3582 at p. 187, and 33 30 copy 11 at fol. 237r; Kirgate's second copy of the *Description,* 33 30 copy 29; John D. Fry's copy of the *Description,* 33 30 copy 30; and Mrs. Damer's Scrapbook, folio 53 D18 828 V.1; there is also a loose copy in proof state, before the publication line was added.

Reproduced: Paget Toynbee, ed., *Journal of the Printing-Office at Strawberry Hill* (New York: Houghton Mifflin Company, 1923), 20; Paget Toynbee, ed., *Strawberry Hill Accounts: A Record of Expenditure in Building Furnishing, &c. Kept by Mr Horace Walpole from 1747 to 1795* (Oxford: Clarendon Press, 1927), 56; Bamber Gascoigne and Jonathan Ditchburn, *Images of Twickenham* (Richmond upon Thames: Saint Helena Press, 1981), Gallery 317; and

Nicolas Barker, ed., *Horace Walpole's Description of the Villa at Strawberry-Hill: A facsimile of the copy extra-illustrated for Charles Bedford in the collection of Lord Waldegrave of North Hill* (Privately printed: The Roxburghe Club, 2010).

Exhibition History: *Horace Walpole and Strawberry Hill*, Orleans House Gallery, Twickenham, 1980, no. 225.

The same view as Catalogue 1.3, 1.4, and 1.6.

{ 1.6 ❧ fig. 27 }

“FARM YARD AND PRINTING HOUSE BELONGING TO STRAWBERRY HILL”
Artist unknown, after Edward Edwards (1738–1806), after 1783–84
Pen and ink and watercolor on wove paper, 18.3 x 23.9 (image), 21.3 x 28.5 (sheet)
Title inscribed in pen and ink below image
The Lewis Walpole Library, Yale University

The same view as Catalogue 1.3, 1.4, and 1.5.

{ 1.7 ❧ fig. 30 }

PRINTING HOUSE AT STRAWBERRY HILL
Philip Audinet (1766–1837), after Edward Edwards (1738–1806), 1817
Etching and engraving on wove paper, vignette, 9.8 x 10.5 (image)
Inscribed in the plate: “E. Edwards del P. Audinel [sic] Sculp”
From the Reverend Thomas Frognall Dibdin, *The Bibliographical Decameron; or, Ten Days Pleasant Discourse upon Illuminated Manuscripts, and Subjects Connected with Early Engraving, Typography, and Bibliography* (London, 1817), 3: 450
The Lewis Walpole Library, Yale University, 113 D54 817

Reproduced: Bamber Gascoigne and Jonathan Ditchburn, *Images of Twickenham* (Richmond upon Thames: Saint Helena Press, 1981), Gallery 318.

79 (Cat. 1.9)

{ 1.8 ❧ fig. 31 }

“THE PRINTING HOUSE OF HORACE WALPOLE AT STRAWBERRY HILL”
William Collard (1792–1847), after Philip Audinet (1766–1837), ca. 1820s
Steel engraving mounted on card, vignette, 9.8 x 14.9 (plate)
Inscribed in the plate: “The Printing House of Horace Walpole at Strawberry Hill. / Engraved by William Collard from an Original Picture.”
Portrait File, Kirgate folder
The Lewis Walpole Library, Yale University

A smaller version of Catalogue 1.7.

{ 1.9 ❧ fig. 79 }

“THE PRINTING OFFICE AT STRAWBERRY-HILL”
Artist unknown, ca. 1930
Zinc line cut in double-ruled frame on paper, 5.4 x 6.3
From *Horace Walpole and His Printing Office at Strawberry Hill* (New Rochelle: Walpole Printing Office, 1930), ix
The Lewis Walpole Library, Yale University, 486 16

{ 1.10 ঌ fig. 43 }

STRAWBERRY HILL, FARMYARD, AND PRINTING HOUSE FROM THE COMMON FIELD
Artist unknown, n.d., but after erection of Beauclerk Tower in 1776 and before erection of Offices in 1790
Pen and ink and watercolor on laid paper, 21.8 x 30.9
The Friends of Strawberry Hill

Reproduced: *Serendipity,* Newsletter 26 of the Friends of Strawberry Hill (Summer 2009): (1).

View looking south, with the Round Tower of Strawberry Hill center left, and the farmyard buildings to the right, showing at the far right the north entrance facade and the west side elevation of the Printing House.

{ 1.11 ঌ fig. 45 }

"NORTHEAST VIEW OF THE PRINTING-HOUSE, AT STRAWBERRY-HILL."
John Carter (1748–1817), probably mid-1780s
Pencil on paper, 21.2 x 18.8 (irreg.)
Title inscribed in pen and ink in Carter's hand, below image
Laid onto fol. 156v of Carter's extra-illustrated copy of the *Description,* 130368
The Huntington Library, Art Collections, and Botanical Gardens

Taken from the same perspective as the Edwards view, Catalogue 1.3–1.6, but not showing the barn on the left.

{ 1.12 ঌ fig. 48 }

INTERIOR OF THE PRINTING HOUSE
John Carter (1748–1817), probably 1785
Pencil on paper, 13.7 x 23.4 (irreg.)
Laid (upside down) into fol. 23v of Carter's extra-illustrated copy of the *Description,* 130368
The Huntington Library, Art Collections, and Botanical Gardens

Outline preparatory sketch for Catalogue 1.14.

{ 1.13 ঌ fig. 49 }

THREE FIGURE SKETCHES INSIDE THE PRINTING HOUSE
John Carter (1748–1817), probably 1785
Pencil on paper, 20.4 x 15.4 (single sheet; irreg.)
Laid onto fol. 158v of Carter's extra-illustrated copy of the *Description,* 130368
The Huntington Library, Art Collections, and Botanical Gardens

{ 1.13.1 }
THOMAS KIRGATE COMPOSING AT THE FRAME
Kirgate setting type with window over his right shoulder, a preparatory sketch for Catalogue 1.14.

{ 1.13.2 }
PRINTER AT WORK AT THE PRESS
Printer seen from behind working at the press, a preparatory sketch for Catalogue 1.14.

{ 1.13.3 }
KIRGATE AND TWO YOUNG MEN SEATED AT A TABLE

{ 1.14 ঌ fig. 46 }

"INSIDE OF THE PRINTING-HOUSE AT STRAWBERRY-HILL."
John Carter (1748–1817), probably 1785
Pen and ink and watercolor with wash-line mount on laid paper, 14 x 21.9 (image), 20.7 x 28.8 (mount)
Title inscribed in pen and ink on mount
Laid into Richard Bull's extra-illustrated copy of the *Description,* 33 30 copy 11, at fol. 240r
The Lewis Walpole Library, Yale University

Reproduced: A.T. Hazen, *A Bibliography of the Strawberry Hill Press* (New Haven: Yale University Press, 1942), frontispiece; W.S. Lewis, *Horace Walpole: The A. W. Mellon Lectures in the Fine Arts 1960* (New York: Pantheon Books, 1960; London: Rupert Hart-Davis, 1961), 140; *The Yale Edition of the Correspondence of Horace Walpole* 40 (New Haven and London: Yale University Press, 1980), 117; John Cornforth, "The Cowles-Lewis House, Farmington—II," *Country Life* 63, no. 4217 (4 May 1978): 1233; and Michael Snodin, with the assistance of Cynthia Roman, ed., *Horace Walpole's Strawberry Hill* (New Haven and London: Lewis Walpole Library, Yale Center for British Art, and Victoria and Albert Museum, in association with Yale University Press, 2009), 243, fig. 274.

The principal view of the interior of the Printing House, looking north in the composing room toward the entrance front, Kirgate setting type, with his assistant visible at the press through the door to the right.

{ 1.15 ❧ fig. 47 }

"INTERIOR OF THE PRINTING HOUSE AT STRAWBERRY HILL, WITH A PORTRAIT OF MR. KIRGATE THE WORTHY PRINTER EMPLOY'D BY LORD ORFORD AT THE TIME OF HIS DEATH, AND FOR MANY YEARS PREVIOUS THERETO."
John Carter (1748–1817), 1785 or later
Pen and ink and watercolor with ruled edging in red and black on paper, 15.6 x 22.1 (image)
Title inscribed in pen and ink below image and above swelled rule, after 1797
Laid in at fol. 14 of Richard Bull's *Strawberry Hill Detached Pieces,* 90524
The Huntington Library, Art Collections, and Botanical Gardens

An alternative version of Catalogue 1.14, with variations.

{ 1.16 ❧ fig. 54 }

INTERIOR OF THE PRINTING HOUSE
Attributed to George Perfect Harding (1779/80–1853), ca. 1800
Pen and ink and watercolor with ruled edging on wove paper, 14.5 x 21
Laid into Kirgate's elephant folio collection of Strawberry Hill images
The Lord Waldegrave of North Hill

{ 1.17 ❧ fig. 64 }

EXTERIOR OF THE PRINTING HOUSE
Attributed to George Perfect Harding (1779/80–1853), ca. 1790s
Pen and ink and watercolor with ruled edging on laid paper, 18.5 x 23
Laid into Kirgate's elephant folio collection of Strawberry Hill images
The Lord Waldegrave of North Hill

{ 1.18 ❧ fig. 29 }

"VIEW FROM THE CHAMBER WINDOW OF THE PRINTING HOUSE AT STRAWBERRY HILL"
Silvester Harding (1745–1809), probably 1790s
Pen and ink and watercolor with wash-line mount on wove paper, 12.9 x 9.5 (image), 16.1 x 12.8 (mount)
Signed in pencil: "S.H."
Title inscribed on verso in pencil in Kirgate's hand (part of inscription trimmed away); a note by W.S. Lewis authenticates Kirgate's hand.
The Lewis Walpole Library, Yale University, 49.3678

The view is to the west, from one of the windows on the west side of Kirgate's chamber, to the fields beyond the Teddington road. The same view as Catalogue 1.19, the one copied from the other; it is impossible to determine which of the two is the original. Unlike Catalogue 1.19, this watercolor has faded.

{ 1.19 ❧ fig. 80 }

"VIEW FROM THE PRINTING HOUSE"
George Perfect Harding (1779/80–1853), probably 1790s
Pencil and watercolor with wash-line mount on wove paper, 12.8 x 9.4 (image), 16 x 12.6 (mount)
Signed in pen and ink: "G.P.H."
Title inscribed in pencil in Kirgate's hand, below mount
Laid into Kirgate's elephant folio collection of Strawberry Hill images
The Lord Waldegrave of North Hill

The same view as Catalogue 1.18, the one copied from the other; it is impossible to determine which of the two is the original.

80 (Cat. 1.19)

{ 1.20 ❧ fig. 65 }

"PRINTING-HOUSE AT STRAWBERRY-HILL"
George Perfect Harding (1779/80–1853), after William Bawtree (1744–1824), ca. 1790s
Etching on wove paper, 8.5 x 11.6 (image), 10.5 x 14.3 (plate)
Inscribed in pencil in Kirgate's hand below image: "W. Bawtree del. G. Harding sc. / Printing-house at Strawberry-hill"
Northwick Park collection of Loose Pieces, 336J
The Lewis Walpole Library, Yale University

Another copy is bound in as a frontispiece to Edward Astley's extra-illustrated copy of the *Miscellaneous Antiquities* and *Seven Original Letters of Edward VI,* 335A.

Reproduced: W.S. Lewis, ed., *A Selection of the Letters of Horace Walpole* (New York: Harper & Brothers; London: Oxford University Press, 1926), 1: 191.

View of the north entrance facade of the Printing House; generally similar perspective to Catalogue 1.17 and 1.21.

{ 1.21 ❧ fig. 66 }

"PRINTING OFFICE AT STRAWBERRY-HILL"
Artist unknown, ca. 1810
Pen and ink and watercolor on paper, 7 x 10 (max.), the image cut out around the outline of the trees behind the building
Title inscribed in pencil below image
Laid onto the front free endpaper of Thomas Gosden's holograph "List of Pieces Printed at Strawberry Hill taken from Baker's List," with Gosden's engraved pictorial bookplate (later owned by Lewis Buddy III, who added his own pictorial bookplate), 486 13 quarto
The Lewis Walpole Library, Yale University

{ 1.22 ❧ fig. 67 }

"THE PRINTING OFFICE ERECTED BY HORACE WALPOLE AT STRAWBERRY HILL"
Artist unknown, early 1820s
Pen and ink and wash on wove paper, 17.8 x 25 (image), 23.5 x 30.7 (sheet)
Title inscribed in pencil below image
The Lewis Walpole Library, Yale University, 49 3678

{ 1.23 ❧ fig. 69 }

THE STRAWBERRY HILL PRINTING OFFICE
Artist unknown, 19th century
Pen and ink on paper, mounted on card, 10.4 x 18.8 (sheet)
Inscribed on the verso: "Lewis Buddy III. Feb. 1946 (E.C. Buddy). The SH Printing Office."
The Lewis Walpole Library, Yale University

Reproduced: Munson Aldrich Havens, *Horace Walpole and the Strawberry Hill Press 1757–1789* (Canton, Pa.: Lewis Buddy 3rd, The Kirgate Press, 1901), 48.

An imaginative reconstruction, reproducing the appearance of the original (which had been reduced and restructured more than twenty-five years earlier) but compressing it, and suggesting wooden boarding rather than brick.

{ 1.24 ❧ figs. 74–75 }

PHOTOGRAPHS OF THE FORMER PRINTING HOUSE
Seven gelatin silver photographic prints on paper, ca. 1900
Strawberry Hill Printing House file
The Lewis Walpole Library, Yale University

{ 1.24.1 ❧ fig. 74 }
View of north facade as reconstructed, 17.8 x 12.8 (two copies)

{ 1.24.2 }
Almost identical view of north facade, 16.3 x 11.1

{ 1.24.3 }
Additional almost identical view of north facade, laid on card, 14 x 10.9
Inscribed in pencil on verso: "Ex dono Lewis Buddy II [sic] Printing Office circ. 1900."

{ 1.24.4 }
View of north facade with man standing at door, laid on card, 13.9 x 8.2
Inscribed in pencil on verso by W. S. Lewis: "From Lewis Buddy III Taken at SH c. 1900"

{ 1.24.5 ❧ fig. 75 }
View of north and east facades, showing stable block to the south, 12.8 x 17.7 (two copies)

{ 1.24.6 }
Almost identical view of north and east facades, 11 x 16.2

{ 1.24.7 }
Additional almost identical view of north and east facades, 10.7 x 14.1
Inscribed in pencil on verso: "? Ex dono Lewis Buddy, c. 1900."

Section Two: Maps of the Strawberry Hill Estate Showing the Printing House

{ 2.1 ❧ fig. 44 }

"A MAP OF THE MANOR OF ISLEWORTH IN THE COUNTY OF MIDDLESEX BELONGING TO HIS GRACE THE DUKE OF NORTHUMBERLAND TAKEN FROM AN ACTUAL SURVEY, CORRECTED & DRAWN IN THE YEAR 1786 AND 1787 BY C.J. SAUTHIER."
Claude Joseph Sauthier (1736–1802), 1786–87
Pen and ink and watercolor on paper backed onto linen, 280 x 231
Collection of the Duke of Northumberland, Sy:B.XIII.1e

{ 2.2 ❧ fig. 11 }

"[SLIGHT SKETCH OF THE] GENERAL GROUND PLOTT OF THE GOTHIC MANSION, AND THE GROUNDS ADJACENT, AT STRAWBERRY-HILL."
John Carter (1748–1817), ca. 1790–91
Watercolor with pen and ink on laid paper, 46 x 31
Laid onto verso of front free endpaper of Walpole's extra-illustrated copy of the *Description,* 49 3582
The Lewis Walpole Library, Yale University

Reproduced: W. S. Lewis, "The Genesis of Strawberry Hill," *Metropolitan Museum Studies* 5, part 1 (1934): 85; Isabel Wakelin Urban Chase, *Horace Walpole: Gardenist. An Edition of Walpole's "The History of the Modern Taste in Gardening" with an Estimate of Walpole's Contribution to Landscape Architecture* (Princeton: Princeton University Press, 1943), facing p. 2; and Michael Snodin, with the assistance of Cynthia Roman, ed., *Horace Walpole's Strawberry Hill* (New Haven and London: Lewis Walpole Library, Yale Center for British Art, and Victoria and Albert Museum, in association with Yale University Press, 2009), 23, fig. 29.

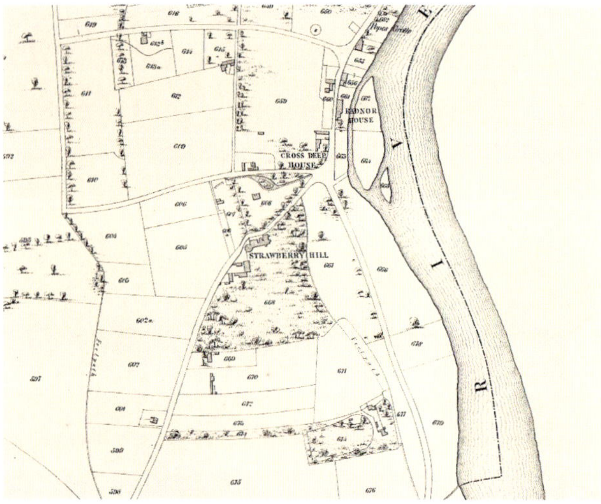

81 (Cat. 2.6; detail)

{ 2.3 fig. 12 }

"SLIGHT SKETCH OF THE GROUND-PLOT OF STRAWBERRY-HILL."
John Carter (1748–1817), ca. 1790–91
Watercolor with pen and ink and ruled edging on laid paper, 31 x 29
Title and key inscribed in pen and ink below the drawing, on the bound leaf
Laid into fol. 23r of Richard Bull's extra-illustrated copy of the *Description,* 33 30 copy 11
The Lewis Walpole Library, Yale University

Reproduced: *The Shell Seat. Strawberry Hill* (Taddington: Architectural Heritage, 2009), 9.

A version of Catalogue 2.2, with different lettering, the Printing House being here marked "E" (rather than "D").

{ 2.4 fig. 14 }

HARDING'S ESTATE MAP OF STRAWBERRY HILL
George Perfect Harding (1779/80–1853), 1801
Pen and ink on wove paper, 27.7 x 27.7
Laid into Kirgate's elephant folio collection of Strawberry Hill images
The Lord Waldegrave of North Hill

Shows the Offices, the Printing House (though no longer identified as such in the Reference), and, set back slightly to the south, the ice house erected in 1795.

{ 2.5 fig. 68 }

TWICKENHAM ENCLOSURE AWARD MAP: PLAN F
1818
Pen and ink and wash on parchment, 67 x 84
Bound into the Enclosure Award, a manuscript folio bound in old calf
Local Studies Library, London Borough of Richmond upon Thames

Shows the former Printing House in its truncated form (square rather than rectangular), consistent with the wash drawing of the early 1820s (Catalogue 1.22). The ice house is also clearly shown.

{ 2.6 fig. 81 }

"PLAN OF THE PARISH OF TWICKENHAM MIDDLESEX"
William Thomas Warren, Surveyor, Isleworth, 1846
Zincograph on paper backed onto linen, 76 x 128
Printed by C.F. Cheffins Zincog. Southampton Buildings, Holborn, London
Collection of the Duke of Northumberland, SyA:P.II.72

There is also a photocopy at the Local Studies Library, London Borough of Richmond upon Thames.

{ 2.7 fig. 71 }

ORDNANCE SURVEY 25" MAP OF MIDDLESEX
Surveyed in 1863
Sheet XX:15, zincograph on paper, 69 x 101
Local Studies Library, London Borough of Richmond upon Thames

Shows Lady Waldegrave's alterations to Walpole's former farmyard, creating a new stable yard, U-shaped in plan and oriented to open to the Teddington road on its west. The former Printing House forms the north end of the easternmost stable building, and is shown as slightly wider than the stable building's southern extension.

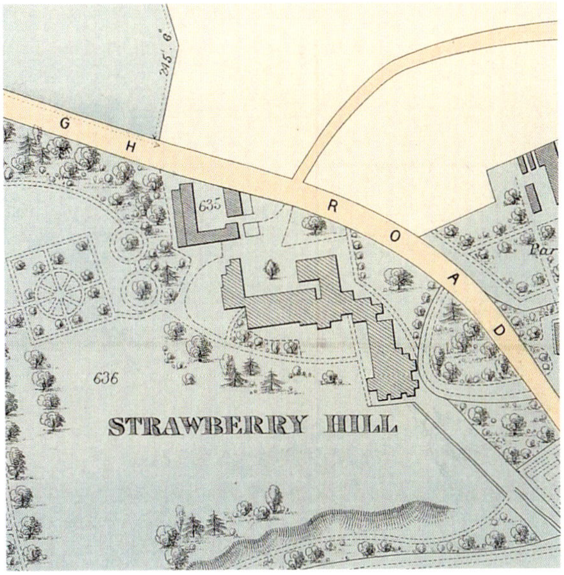

82 (Cat. 2.10; detail)

{ 2.8 ☙ fig. 72 }

"PLAN OF WALDEGRAVE PARK ESTATE. TWICKENHAM. TO BE SOLD OR LET ON BUILDING LEASES."
F.T.R., ca. 1880
Hand-colored lithograph on wove paper, 21 x 26.3
Signed in the stone: "F.T.R. fecit"
Printed at Standidge & Co. Litho, 36 Old Jewry FCA Moorfields
The Lewis Walpole Library, Yale University

Shows Strawberry Hill Road and Waldegrave Park Gardens laid out as building plots, as part of Lord Carlingford's initial disposal of part of the Estate; with a vignette of a "Design for Villa" and a note referring potential buyers to F. Nesbitt Kemp, Architect and Surveyor, or J.R. Macarthur Esq., Solicitor. Lady Waldegrave's stable block is visible in essentially the same format as in Catalogue 2.7.

{ 2.9 ☙ fig. 73 }

"THE STRAWBERRY HILL ESTATE TWICKENHAM. FOR SALE BY MESSRS. VENTOM, BULL & COOPER,"
1883
Hand-colored lithograph on glazed paper, 88 x 59
Printed by Waterlow Bros. & Layton, Birchin Lane, London
Bound in at the end of the Auction Catalogue for the sale of the Residential Estate of Strawberry Hill, 15 June, 1883
The Lewis Walpole Library, Yale University

The Lewis Walpole Library has three additional maps of the Estate, none of them dated, but each referring to Lord Carlingford's sale of the Estate.

Shows Lady Waldegrave's U-shaped stable yard as in Catalogue 2.7, but with the map oriented toward the west. The truncated remains of the Printing House are visible at the bottom right (northeast) corner of the stable yard.

{ 2.10 ☙ fig. 82 }

"THE STRAWBERRY HILL ESTATE, FOR SALE BY AUCTION ON TUESDAY 10TH JULY 1888. BY MESSRS. HAMPTON & SONS."
1888
Hand-colored lithograph on wove paper, 56.5 x 43
Printed by T. Pettitt & Co. Lith 29, Frith St. Soho W
Bound in at the end of the Auction Sale Catalogue of the Strawberry Hill Estate, on the instructions of the trustees of the late Baron de Stern
The Lewis Walpole Library, Yale University, Folio 485 888 st 4

There is another copy at the Local Studies Library, London Borough of Richmond upon Thames.

In relation to the site of the Printing House and stables, this replicates the map in Catalogue 2.9.

Section Three: Portraits of Thomas Kirgate

{ 3.1 ❧ fig. 36 }

"T.K." [THOMAS KIRGATE]
Edward Edwards (1738–1806), finished by John Hall (1739–1797), 1784
Etching on laid paper, 12.5 x 8.3 (plate)
Inscribed in the plate: "Painted & etch'd by E.E. Finished by I. Hall. / T.K."
Inscribed lightly in the plate at upper left: "47"
Inscribed in pencil in Kirgate's hand, below image: "The best likeness. T.K."; Kirgate has also, in pencil, completed his last name after the etched initial "K"
From Kirgate's second extra-illustrated copy of the *Description,* 33 30 copy 29
The Lewis Walpole Library, Yale University

Copies of the proof state of the print, before letters and without the number 47, are in Walpole's manuscript "Journal of the Printing Office at Strawberry-hill near Twickenham in Middlesex," 49 2506A, laid onto p. 18, inscribed by Walpole, "E. Edwards p. & sc. 1784 Th: Kirgate Printer at Strawberry hill"; *A Collection of all the Loose Pieces Printed at Strawberry-Hill,* 49 2506, following the title page, annotated by Walpole "Thomas Kirgate Printer at Strawberry hill. painted and etched by Edw. Edwards, finished in the Engraving by J. Hall 1784," with pencil note in Kirgate's hand "Aetat 49"; in the Portrait File, Kirgate folder, annotated by Kirgate in pencil supplying the missing text; at p. 187 of Walpole's copy of the *Description,* 49 3582, inscribed in pen and ink beneath print in Walpole's hand "T. Kirgate the Printer."; at fol. 240r of Richard Bull's extra-illustrated copy of the *Description,* 33 30 copy 11, with "Mr. Walpole's Printer" written beneath; tipped in before the engraved title page of Edward Edwards, *A Collection of Views and Studies after Nature with Other Subjects Designed and Etched by Edward Edwards, Associate and Teacher of Perspective in the Royal Academy* (London, 1790), annotated in pencil by Edwards "Thomas Kirgate Printer to the Hon H Walpole Painted & etched by E Edwards Finished by J Hall," 75 Ed 96 790; and another copy annotated by Kirgate in pen and ink over pencil "Thomas Kirgate, Printer at Strawberry-hill. Painted & Etched by E. Edwards 1784. Finished by J. Hall." There are other proofs before letters at the Huntington Library, in Richard Bull's volume of *Strawberry Hill Detached Pieces,* 90524, at fol. 13, inscribed by hand "T. Kirgate Printer at Strawberry Hill… E. Edwards pinxt. & fecit 1784"; and immediately preceding the Preface in the *Description,* 131207.

The print as lettered and with the number 47 in the upper left corner was published by Edwards as the forty-seventh of fifty-one plates in *A Collection of Views.* Separate copies of the print with lettering at the Lewis Walpole Library are in volume 4 of Thomas Mackinlay's extended and extra-illustrated copy of the Strawberry Hill sale catalogue, *A Catalogue of the Classic Contents of Strawberry Hill Collected by Horace Walpole* (1842), 485 842 C76 IV copy 3. Other copies are in Kirgate's elephant folio collection of Strawberry Hill images (Lord Waldegrave of North Hill); and at the British Museum, as recorded in Freeman O'Donoghue, *Catalogue of Engraved British Portraits Preserved in the Department of Prints and Drawings in the British Museum* (London: British Museum, 1910), 2: 705, item 1. See also the entry in Henry Bromley, *A Catalogue of Engraved British Portraits* (London, 1793), 406.

Reproduced: Paget Toynbee, ed., *Journal of the Printing-Office at Strawberry Hill* (New York: Houghton Mifflin Company, 1923), 48.

Exhibition History: *Horace Walpole and Strawberry Hill,* Orleans House Gallery, Twickenham, 1980, no. 224.

Kirgate, quarter-length, looking to right. After an untraced portrait exhibited by Edwards at the Royal Academy in 1784, no. 152.

{ 3.2 ❧ fig. 37 }

"THOS. KIRGATE LORD ORFORD'S PRINTER AT STRAWBERRY HILL."
G. Barrett, after Edward Edwards (1738–1806), after 1791
Etching on wove paper, 12 x 7.6 (sheet)
Inscribed in the plate: "Lord Orford's Printer at Strawberry Hill / Thos. Kirgate [an etched facsimile of Kirgate's signature] / Pubd. by G. Barrett Engraver 289 Holborn."
Portrait File, Kirgate folder
The Lewis Walpole Library, Yale University

There is an additional copy at the Lewis Walpole Library, laid in to face p. 177 of vol. 2 of L.B. Seeley, *Horace Walpole and His World* (1884), 225 884s copy 5, but with the publication line cropped. Another copy is in the British Museum, registration no. 1876, 1209.325, as recorded in Freeman O'Donoghue, *Catalogue of Engraved British Portraits Preserved in the Department of Prints and Drawings in the British Museum* (London: British Museum, 1910), 2: 705, item 2; this item appears also to be the second print of Kirgate noted in Henry Bromley, *A Catalogue of Engraved British Portraits* (London, 1793), 406.

Reproduced: Nicolas Barker, ed., *Horace Walpole's Description of the Villa at Strawberry-Hill: A facsimile of the copy extra-illustrated for Charles Bedford in the collection of Lord Waldegrave of North Hill* (Privately printed: The Roxburghe Club, 2010), following p. 96 of facsimile, as proof before letters.

Based on Catalogue 3.1.

{ 3.3 ❧ fig. 38 }

"THOMAS KIRGATE PRINTER TO HORACE WALPOLE AT STRAWBERRY HILL."

William Collard (1792–1847), after G. Barrett, ca. 1820s

Steel engraving on chine-collé, 20.7 x 13.8 (image), 24 x 14.8 (plate)

Inscribed in the plate: "Thomas Kirgate / Printer to / Horace Walpole at Strawberry Hill. / Engraved by William Collard from an Original Picture."

Portrait File, Kirgate folder

The Lewis Walpole Library, Yale University

Based on Catalogue 3.2.

{ 3.4 ❧ fig. 41 }

"THOMAS KIRGATE, PRINTER AT STRAWBERRY=HILL. WITH A VIEW OF THE PRINTING HOUSE."

Andrew Birrell (flourished 1769–1820), after Silvester Harding (1745–1809), 1798

Etching, engraving, and stipple on laid paper, 12.4 x 11 (image), 16.5 x 13.3 (plate)

Inscribed in the plate: "S. Harding ad vivum delin. 1794. A. Birrell sculp. / Thomas Kirgate, / Printer at Strawberry=Hill. / With a View of the Printing House. / Published in 1798."

Laid into Richard Bull's extra-illustrated copy of the *Description,* 33 30 copy 11, fol. 237r

The Lewis Walpole Library, Yale University

This image was much used in extra-illustration. There are examples in the Portrait File, Kirgate folder; and in the Northwick Park collection of Loose Pieces, 336J, which also has an early proof before letters, inscribed in pencil "but 2 impressions taken" (see Fig. 42). There are further copies laid into Kirgate's copy of the *Description,* 33 30 copy 6, leaf immediately following title page; in his second copy, 33 30 copy 29, following p. 86; and in John D. Fry's copy of the *Description,* 33 30 copy 30. A copy is also laid in (with Kirgate's signature beneath it, "T. Kirgate, Strawberry Hill") as a frontispiece to the extra-illustrated copy of the *Description* at the Beinecke Rare Book and Manuscript Library, Yale University, JAX 099.

Reproduced: W.S. Lewis, ed., *The Yale Edition of Horace Walpole's Correspondence* 41 (New Haven: Yale University Press, 1980), 283; Stephen Calloway, Michael Snodin, and Clive Wainwright, *Horace Walpole and Strawberry Hill,* exh. cat. (Richmond: London Borough of Richmond upon Thames, 1980), 70; Bamber Gascoigne and Jonathan Ditchburn, *Images of Twickenham* (Richmond upon Thames: Saint Helena Press, 1981), Gallery 319; Michael Snodin, with the assistance of Cynthia Roman, ed., *Horace Walpole's Strawberry Hill* (New Haven and London: Lewis Walpole Library, Yale Center for British Art, and Victoria and Albert Museum, in association with Yale University Press, 2009), 243, fig. 273; and Nicolas Barker, ed., *Horace Walpole's Description of the Villa at Strawberry-Hill: A facsimile of the copy extra-illustrated for Charles Bedford in the collection of Lord Waldegrave of North Hill* (Privately printed: The Roxburghe Club, 2010), facing p. 96 of facsimile.

Exhibition History: *Horace Walpole and Strawberry Hill,* Orleans House Gallery, Twickenham, 1980, no. 226.

Kirgate in profile with the Printing House behind him, to the right, showing all but the west end of the north entrance front, and part of the front garden.

83 (Cat. 3.5)

{ 3.5 ❧ fig. 83 }

"THOMAS KIRGATE. PRINTER AT STRAWBERRY=HILL."
David Edward Cronin (1839–1925), after Andrew Birrell (flourished 1769–1820), probably 1880s
Pen and ink on wove paper, 12.4 x 11 (to ruled line), with below it a watercolor vignette of a hanging shelf of books
Signed at lower right, within image: "D.E.C."
Title inscribed in pen and ink below the image
Bound into p. 196 of the extra-illustrated and extended copy of L.B. Seeley, *Horace Walpole and His World* (1884), 225 884S, copy 2, vol. 1.
The Lewis Walpole Library, Yale University

The drawing is a copy of Catalogue 3.4.

{ 3.6 ❧ fig. 50 }

THOMAS KIRGATE
Silvester Harding (1745–1809), 1790s
Pencil and watercolor with wash-line mount on wove paper, 16 x 13.2 (image), 21 x 18.1 (mount), 36 x 25.6 (sheet)
Inscribed in pen in Kirgate's hand, beneath the mount: the signature "Thos. Kirgate"
The Trustees of the British Museum, regn. no. 1867, 0309.1846

Bibliography: Freeman O'Donoghue, *Catalogue of Engraved British Portraits Preserved in the Department of Prints and Drawings in the British Museum* (London: British Museum, 1910), 2: 705, item 3.

Exhibition History: *Horace Walpole and Strawberry Hill,* Orleans House Gallery, Twickenham, 1980, no. 223.

Three-quarter length, standing in front of his compositor's table, holding his composing stick in his left hand. On the composing stick are (upside-down) the letters "ANECDOTES." The sheet of paper laid on the upper case of the compositor's table is blank.

{ 3.7 ❧ fig. 51 }

"THOS. KIRGATE. PRINTER AT STRAWBERRY HILL."
Edward Harding (1771/6–1791/6), after Silvester Harding (1745–1809), n.d., but no later than 1796 (date of death of Edward Harding junr., according to the *Oxford Dictionary of National Biography* entry for his father, Silvester Harding)
Etching and stipple engraving on wove paper, 16.8 x 13.9 (image), 22.5 x 16.7 (plate)
Inscribed in the plate: "S. Harding delt. E. Harding Junr. Sculp. / Thos. Kirgate. / Printer at Strawberry Hill."
Inscribed in pencil in Kirgate's hand: "The Plate destroyed, being unlike the Person"; and below, "For George Baker, Esqr. with Kirgate's Respects"
Northwick Park collection of Loose Pieces, 336J
The Lewis Walpole Library, Yale University

Another copy (but with the sheet trimmed so as to excise the line "Printer at Strawberry Hill") laid into Kirgate's elephant folio collection of Strawberry Hill images (The Lord Waldegrave of North Hill).

In addition to the composing stick showing the letters "ANECDOTES," the sheet of paper laid on the upper case of the compositor's table is lettered as a version of the title page of Volume I of the *Anecdotes of Painting* (which Kirgate did not print).

{ 3.8 ❧ fig. 84 }

"THOMAS KIRGATE. PRINTER AT STRAWBERRY HILL."
Early 20th century
Photogravure on wove paper, 15.8 x 12.9 (image), 21.4 x 15.8 (plate)
Portrait File, Kirgate folder
The Lewis Walpole Library, Yale University

From a now lost state of Catalogue 3.7, before lettering on both the sheet of paper on the upper case and the composing stick.

84 (Cat. 3.8)

{ 3.9 ❧ fig. 62 }

"MR. KIRGATE, LATE PRINTER AT STRAWBERRY HILL FOR 30 YEARS. DRAWN BY T. GOSDEN FROM AN ORIGINAL DRAWING."
Thomas Gosden (1780–1843), 1800–10
Pencil on laid paper, 7.2 x 6.5 (portrait), beneath a sketched festoon and above a copy of the Strawberry Hill larger vignette
Laid onto the verso of the front free endpaper of Thomas Gosden's holograph "List of Pieces Printed at Strawberry Hill, taken from Baker's List," with Gosden's engraved pictorial bookplate (later owned by Lewis Buddy III, who added his own pictorial bookplate), 486 13 quarto
The Lewis Walpole Library, Yale University

Reproduced: Munson Aldrich Havens, *Horace Walpole and the Strawberry Hill Press 1757–1789* (Canton, Pa.: Lewis Buddy 3rd, The Kirgate Press, 1901), 48.

Kirgate as an elderly man, head and shoulders in right profile. See note 175 for a possible untraced print of Kirgate that may be related to this image.

{ 3.10 ❧ fig. 63 }

"THOMAS KIRGATE MANY YEARS PRINTER TO HORACE WALPOLE EARL OF ORFORD AT STRAWBERRY HILL. DIED JUNE 11 1810 AGED 76."
George Perfect Harding (1779/80–1853), 1810
Watercolor on paper, 12.4 x 10 (image, including painted frame)
Signed and dated in pen and ink along left edge of oval frame: "GP Harding delt. 1810"
Title inscribed in pen and ink below image
The Lewis Walpole Library, Yale University

Reproduced: W.S. Lewis, *The Forlorn Printer Being Notes on Horace Walpole's Alleged Neglect of Thomas Kirgate* (Farmington: Privately printed, 1931), frontispiece.

Head and shoulders in oval frame, looking to the left.

Index

Bibliographies and exhibition histories detailed in the Iconography section (pp. 121–34) are not indexed. Italicized page numbers reference illustrations.

EDITED BY LESLEY K. BAIER

DESIGNED AND SET IN TYPE BY JOHN GAMBELL, YALE UNIVERSITY PRINTER

SET IN ADOBE CASLON TYPEFACES DESIGNED BY CAROL TWOMBLY IN 1990 BASED ON FONTS REPRESENTED IN CASLON TYPE FOUNDRY SPECIMENS ISSUED BETWEEN 1734 AND 1770

PRINTED BY CS GRAPHICS, SINGAPORE

DISTRIBUTED BY YALE UNIVERSITY PRESS